ELECTRICAL EQUIPMENT MANUAL

ELECTRICAL EQUIPMENT MANUAL

Joseph F. McPartland

William J. Novak

Associate Editors, *Electrical Construction and Maintenance*

Third Edition

McGRAW-HILL BOOK COMPANY

New York   San Francisco   Toronto   London   Sydney

ELECTRICAL EQUIPMENT MANUAL

45697

23456789 HL 9876

PREFACE

This book affords a practical, easy-to-use reference for anyone desiring or requiring a ready background in the fundamentals of electrical equipment. The format of carefully balanced copy and simple graphic sketches has been selected for its informality and inviting appearance. The authors sought to relieve the rigid, sophisticated, and often forbidding style which characterizes much technical literature. Throughout, the objective has been to explain in a simple yet comprehensive manner the basic things *you should know* about today's electrical equipment.

The original style and currency of this material should make it of prime interest to a wide range of electrical people: electrical supply salesmen, manufacturers' representatives, electrical contractors, electricians, plant electrical personnel, electrical engineers, electrical utility personnel, architects, electrical estimators, electrical inspectors, electrical draftsmen, and particularly apprentices and students, whose textbooks invariably neglect the practical aspects of the theory studied.

This material first appeared as a continuous series of articles in *Electrical Wholesaling* magazine over the past several years. Its enthusiastic reception by the electrical industry has led to the present compilation in permanent book form.

Joseph F. McPartland
William J. Novak

CONTENTS

Incandescent Lamps

INCANDESCENT lamps operate on a simple principle. An electric current passing through a very high-resistance tungsten wire coil, encased in an evacuated or gas-filled glass bulb, raises the temperature of the coil to an intense "white" heat. The coil is then said to be incandescent, and the white light produced is sufficient for purposes of illumination.

Construction

These are the elements of an incandescent lamp:

Filament—Invariably, the filament is some type of coil of fine tungsten wire. The filament is the light source in the lamp, and its electrical characteristics determine the rating of the lamp. Its wattage (and the wattage of the lamp) equals the voltage applied at the base times the current flowing through it.

Incandescent lamp efficiency is determined by the operating characteristics of the filament. Of the energy put into a filament, only about 12 per cent is converted to light; the balance is converted to heat. As a result, incandescent lamps are inherently inefficient, far less efficient than mercury vapor or fluorescent lamps.

Lamp efficiency is expressed in terms of light produced (lumens) per watt of power consumed. For a given voltage and life rating, high-wattage lamps are more efficient than low-wattage lamps. One 50-watt, 120-volt lamp produces more light than two 25-watt, 120-volt lamps. For a given wattage, low-voltage lamps are more efficient than high-voltage lamps.

Bulb—To prevent disintegration of a filament at incandescent temperature, it must be operated in a vacuum or under pressure of an inert gas, sealed in a glass envelope—the bulb. Soft glass is used for general service lamps; hard glass is used to provide sturdy bulbs for lamps of high operating temperatures—outdoor lamps, projection lamps, spot lamps, etc. Bulb sizes and shapes (shown on opposite page) are determined by lamp application.

For general service lamps, inside frost (etched with acid) is the most common bulb finish. This finish diffuses light output, absorbs very little, and makes the complete surface of the bulb act as the light source. The clear bulb lamp, with the smaller, concentrated filament source of light, is required in optical systems which provide accurate control of a small output source.

Bases—The base provides the means for connecting a lamp to the electric circuit. Types and applications of bases are shown on the opposite page.

Operation

Lamp life and light output are determined by filament temperature. For a given lamp, the higher the filament temperature, the greater the light output and the shorter the lamp life. Light output and lamp life are therefore interdependent.

Some lamps have high output, short life; others, low output, long life. For example, the photoflood lamp with its high light output has a life of about 3 hours; general service lamps have much lower output but last about 800 hours.

General service lamps may be operated in any position, from "base-down" to "base-up." Gas-filled general service lamps (40-watts and up) suffer lower depreciation in output if operated "base-up."

Certain lamps—projection, spot, flood

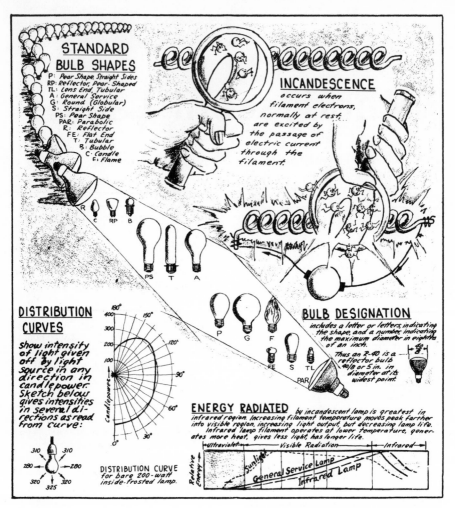

STANDARD BULB SHAPES
P: Pear Shape, Straight Sides
RP: Reflector, Pear-Shaped
TL: Lens End, Tubular
A: General Service
G: Round (Globular)
S: Straight Side
PS: Pear Shape
PAR: Parabolic
R: Reflector
FE: Flat End
T: Tubular
B: Bubble
C: Candle
F: Flame

INCANDESCENCE
occurs when filament electrons, normally at rest, are excited by the passage of electric current through the filament.

DISTRIBUTION CURVES
show intensity of light given off by light source in any direction in candlepower. Sketch below gives intensities in several directions as read from curve:

DISTRIBUTION CURVE for bare 200-watt inside-frosted lamp.

BULB DESIGNATION
includes a letter or letters indicating the shape, and a number indicating the maximum diameter in eighths of an inch.

Thus an R-40 is a reflector bulb 40/8 or 5 in. in diameter at its widest point.

ENERGY RADIATED by incandescent lamp is greatest in infrared region. Increasing filament temperature moves peak farther into visible region, increasing light output, but decreasing lamp life. Infrared lamp filament operates at lower temperature, generates more heat, gives less light, has longer life.

and some series street lamps—must be operated in the position specified by the manufacturer.

Ordinary lamps are highly susceptible to failure if subjected to excessive vibration or shock. For conditions of lamp vibration, vibration service lamps, with specially constructed filaments, should be used. Rough service lamps, with special shock-resisting filaments, are available for applications where excessive shock may be encountered.

Lamp Types

General Service Lamps—Most popular type; find widest range of application.

Watts—6 to 2000.

Bulbs—A, S, G and PS.

Volts—120 (some also in 240-, 60- and 30-volt ratings).

Finishes—clear or inside frosted, 150-watts and up; inside frosted, standard under 150-watts.

Projector Lamps—Two-piece construction of molded, heat resistant, hard glass, with inside aluminum reflector-finish on bulb and a lens cover which determines beam width. Have wide-range outdoor application; cost more and are more efficient than R lamps. Available with spot or flood light distribution.

Sizes—150-watt PAR-38, 200-watt

PAR-46, 300-watt PAR-56; 120-volts.

Reflector Lamps—Mold-blown, soft glass bulb, with aluminum reflector-finish on walls and frosted face, the density of which controls beam width. Should not be exposed to rain or snow or brought into contact with metal parts of equipment. Available in spot or flood types.

Sizes—75-watt R-30, 150-watt R-40, 300-watt R-40, 500-watt R-52, 750-watt R-52; 120-volts.

Silvered Bowl Lamps—A- or PS-shaped bulbs with half of the bowl-end of the bulb coated on the outside with silver. For "base-up" mounting in a variety of suspended or recessed louvered fixtures; provides a range of indirect-type applications. Silver bowl efficiently reflects light up. Sealed reflector remains dirt- and dust-free; lamp requires minimum cleaning.

Sizes—60 to 1000 watts; 120-volts.

Tubular Lamps—T-shaped bulbs with elongated filaments, in clear or frosted finish. For lighting showcases, signs, display boards, and in other applications requiring a linear source of light of minimum diameter, tubular lamps are used. With screw bases, available sizes are 25- and 40-watts (some to 150-watts).

Sealed Beam Lamps—PAR-type lamps for a wide range of low-voltage applications—automotive, marine, aircraft.

Projection Lamps—T-shaped lamps used in picture projection systems. Designed to meet need of precision filament positioning in optical systems lamps have special filament assemblies and pre-focusing bases.

Series Lamps—For street lighting, series lamps are rated in lumens and amperes instead of watts and volts. In a series circuit, all lamps should have same current rating.

Infrared Lamps—Operate on same principles as lamps for lighting. Have lower filament temperature and longer life; produce less light but more infrared energy.

Sizes—125-, 250- and 375-watt R-40; 120-volts.

STANDARD LAMP BASES

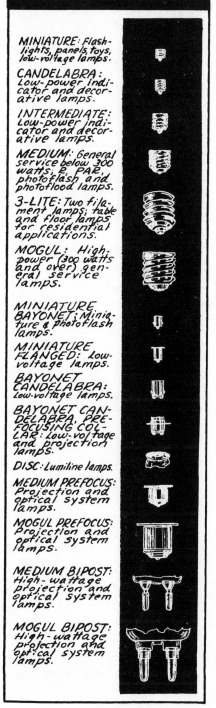

MINIATURE: Flashlights, panels, toys, low-voltage lamps.

CANDELABRA: Low-power indicator and decorative lamps.

INTERMEDIATE: Low-power indicator and decorative lamps.

MEDIUM: General service below 300 watts, R, PAR, photoflash and photoflood lamps.

3-LITE: Two filament lamps; table and floor lamps for residential applications.

MOGUL: High-power (300 watts and over) general service lamps.

MINIATURE BAYONET: Miniature & photoflash lamps.

MINIATURE FLANGED: Low-voltage lamps.

BAYONET CANDELABRA: Low-voltage lamps.

BAYONET CANDELABRA, PRE-FOCUSING COLLAR: Low-voltage and projection lamps.

DISC: Lumiline lamps.

MEDIUM PREFOCUS: Projection and optical system lamps.

MOGUL PREFOCUS: Projection and optical system lamps.

MEDIUM BIPOST: High-wattage projection and optical system lamps.

MOGUL BIPOST: High-wattage projection and optical system lamps.

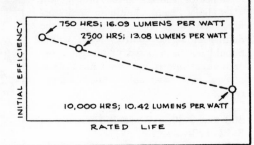
A NUMBER of relatively new characteristics available in incandescent lamps deserve consideration in selection of such lamps:

Inside Silica Coating—Incandescent lamps are given high diffusion of light output by a fine coating of silica on the inside of the bulb. The light output from the filament is distributed evenly over the entire surface of the bulb, eliminating the noticeably bright spot around the filament area of a standard inside-frosted lamp. For applications where the lamp bulb is wholly or partly exposed to view, silica-coated bulbs greatly reduce glare. They are less glaring than standard inside-frosted lamps. Such lamps also soften shadows and minimize the brightness of reflections from specular (shiny) surfaces.

Light output of silica-coated lamps is about the same as that of inside-frosted lamps of the same wattage.

They are available in general purpose bulb shapes up to 150 watts, in modified tubular shape and in the same shape as the R-40 reflector bulb.

Long Life Lamps—Much controversy has developed over the relative merits of the so-called "Long Life" lamp. This lamp has a rated life of, say, 10,-000 hours, compared with a rated life of 750 or 1,000 hours for standard incandescent lamps. Because of its longer life, the long-life lamp has been presented as a superior lamp. But life is not the sole measure of a lamp, and a stronger argument can be made for balanced performance.

Long-life lamps are listed as available in sizes ranging from 15 to 150

THE IODINE CYCLE

In the quartz-iodine lamp, tungsten from the filament adsorbed by the inner surface of the tube reacts with iodine atoms, forming tungsten iodide. The tungsten iodide diffuses back to the filament and subsequently separates again into tungsten and iodine.

ENCLOSING TUBE

FILAMENT

$W + 2I \rightleftarrows WI_2$

This reversible reaction dissipates any accumulation of tungsten on the inside of the tube, eliminating any blackening and thus maintaining a practically constant output throughout the life of the lamp.

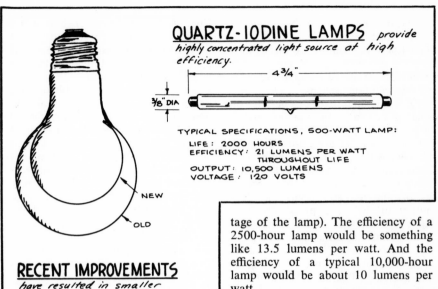

QUARTZ-IODINE LAMPS *provide highly concentrated light source at high efficiency.*

4¾"

⅜" DIA

TYPICAL SPECIFICATIONS, 500-WATT LAMP:

LIFE: 2000 HOURS
EFFICIENCY: 21 LUMENS PER WATT
THROUGHOUT LIFE
OUTPUT: 10,500 LUMENS
VOLTAGE: 120 VOLTS

NEW

OLD

RECENT IMPROVEMENTS
have resulted in smaller physical sizes for given wattage ratings of incandescent lamps.

watts. Such lamps are premium priced compared with standard lamps (3 to 4 times the cost of standard lamps), but they do offer economies in lamp bulb costs, since one 10,000-hour lamp will operate for as long as thirteen 750-hour lamps. For residential applications and for some commercial applications, the long life offered by such lamps is often deemed to be more important than the lower light output and higher operating cost of such a lamp when compared to a standard lamp of the same wattage.

The data of lamp manufacturers shows the lumens of light output from various lamps. Taking lamps of the 100-watt size, for instance, the lumens of light output per watt of energy input varies with the life of the lamp. The longer the life of a lamp, the lower is its lumens per watt or, to put it another way, the lower is its efficiency. In the case of the 100-watt lamps, the standard 750-hour lamp has an efficiency over the life of the lamp of about 16.4 lumens per watt (total lamp lumens divided by the wat-

tage of the lamp). The efficiency of a 2500-hour lamp would be something like 13.5 lumens per watt. And the efficiency of a typical 10,000-hour lamp would be about 10 lumens per watt.

Because the purpose of a light bulb is to produce light efficiently and economically, the manufacturers provide a careful compromise in standard 750 or 1,000-hour incandescent lamps. They are balanced to provide white light efficiently for a reasonable life.

For applications in which longer lamp life is required, the manufacturers make available various "special service" or "extended service" lamps. Typical lamps of this type provide 2,500 hours of operation. Their extended life is achieved with a sacrifice in whiteness of the light and in the efficiency—lumens per watt.

The so-called "long-life" lamp is a continuation of the life vs efficiency modification on the standard lamp. But, again, lumens per watt decreases as life increases.

Quartz-Iodine Lamps—This lamp has gained wide application over recent years for all types of floodlighting tasks. It competes with standard incandescent floodlight lamps in suitable luminaires and with mercury-vapor lighting units.

The quartz-iodine lamp is a tubular type filament lamp with a terminal at each end. It is made in two sizes. The

500-watt lamp is 4½ inches long and is designed to operate at 120 volts. The 1,500-watt lamp is 10 inches long and is designed to operate at 277 volts, for use on the popular 480/277-volt distribution systems.

Quartz-iodine lamps operate on what is known as the "iodine cycle." This is a process by which evaporated tungsten particles are returned to the filament within the tubular bulb. This keeps the inside walls of the bulb completely free of blackening deposits of tungsten and keeps the light output of the lamp practically constant over the life of the lamp. The quartz lamp has an average operating life of 2,000 hours. The quartz lamps will, therefore, last twice as long as the average 1,000-hour 500-watt and 1500-watt incandescent lamps. It has an efficiency of 21 lumens per watt, compared to an average efficiency of 17.5 lumens per watt for the 500-watt PS-40. The 1500-watt quartz lamp has an efficiency of 22 lumens per watt. The most important character-istic of the efficiency of the quartz lamp is that it is constant over the life of the lamp. Standard lamps lose appreciable light output over the operating life of the lamp.

The small size and pencil-like shape of the quartz lamp suit it to use in wrap-around reflector luminaires to produce rectangular patterns of light output. The line-source of light affords accurate control of the light output by the luminaire reflector. The rectangular shape of light output is particularly effective in lighting rectangular areas like large signs, building walls and facades and rectangular parking and driveway areas. Other uses are for airport runways, outdoor amusement parks, used car lots, show windows and industrial areas.

In addition to long life and constant light output over life of the lamp, the quartz lamp is much more resistant to thermal shock than the standard lamp. An operating quartz lamp can even withstand the shock of having ice water poured over it.

Mercury Vapor Lamps

MERCURY vapor lamps are classified as electric discharge lamps—light is produced by gaseous conduction (a passage of electric current through a vapor or gas). A voltage is applied between two electrodes, one at each end within the lamp housing, initiating a flow of current. Interaction between the current electrons and the gas molecules produces light. To limit and stabilize current flow in electric discharge sources, a transformer or similar current-limiting device must be used. The transformer also provides the proper voltage to initiate the arc.

Construction

Mercury lamps generally consist of two bulbs—an arc-tube, containing the electric discharge through the gas, within an outer bulb which protects the arc-tube from the effects of temperature changes. The arc-tube is made of quartz or hard glass; the outer bulb, of hard glass.

Mercury vapor lamps contain vaporized mercury as the "gas." A small amount of argon gas is also used to afford easy starting. It initiates the current flow and vaporizes the small drops of mercury within the lamp. The mercury vapor then conducts the current.

Either tungsten coils with an electron-emissive coating* or elements of thorium metal are used for electrodes.

Energy Output—In the spectrum of light and energy rays, mercury vapor lamps produce ultraviolet rays, visible light rays and infrared rays. By design, mercury lamps vary according to energy output in the spectrum regions. Lamps for lighting have high energy output in the visible regions. Lamps which create fluorescence in dyes and pigments produce energy in the near ultraviolet or "Black Light" region. Other lamps, such as sunlight and vitamin-D producing types and germicidal lamps, produce ultraviolet radiation of shorter wavelengths.

Lamp use determines the choice between quartz or glass for the arc-tube. Quartz arc-tube lamps have better color quality and 25 per cent higher initial light output than glass arc-tube types. The quartz type is smaller, more brilliant and better suited to lens or reflector control of its output. The glass-tube type must be operated in a vertical position; the quartz type, in any position.

Operation

Required starting voltage for a mercury lamp depends upon the electrode arrangement. High voltage (as high as 1200 volts) must be applied between

---

* Easily gives up electrons to produce current.

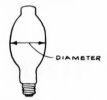
electrodes in two-electrode type lamps. Those lamps with an auxiliary starting electrode, the more common three-electrode lamps, require much lower voltage (220 and 250 volts) for starting.

For each type of mercury lamp, there is a proper transformer to provide required lamp voltage and current-limiting. Although operation of lamp and transformer is ordinarily at low power factor (50-60 per cent), transformers with built-in capacitors are commonly used to raise the power factor to 90 per cent or better. Both single-and two-lamp transformers are available, but wattage loss and stroboscopic effect are minimized with the two-lamp type.

Mercury lamps take several minutes to stabilize to normal operating conditions and maximum light output after initiation of the arc. Interruption of current or a quick decrease of 15 per cent or more in operating voltage will extinguish a lamp.

The number of times a mercury lamp is started greatly affects its light output and life. Each time a lamp is started some of the emissive coating on the electrodes falls away and is deposited on the bulb. The starts continually darken the bulb and ultimately exhaust the emission material. Average life, however, is relatively long.

Temperature of surrounding air has little effect on double-bulb lamps; but single-bulb types are not satisfactory at temperatures below 32° F., unless some protective shield is used. All discharge lamps, however, require higher than normal voltage for starting when used in cold air. Such is the case with outdoor applications in cold climate areas, particularly with quartz arc-tubes.

Very high operating temperatures can do much damage to mercury lamps. Care should be taken that no luminaire used with these lamps traps heat around the lamp.

Applications

True color rendition is not possible with mercury lamps. Although its line spectrum makes the mercury lamp a very efficient source of light, heavy output concentrations in the blue and green regions and absence of sufficient output in the red region distorts colors.

Correction for color distortion can be achieved by using incandescent lamps in combination with the mercury units. To effect noticeable color correction, at least 15 per cent of total light output of a combination of mercury and incandescent lighting should be incandescent. For a ratio greater than 60 per cent incandescent to 40 per cent mercury vapor, negligible color correction is possible by adding more incandescent lighting.

Color corrected mercury lamps, with an inside phosphor coat to add red color, are also available for many applications.

Stroboscopic (flickering) effect of mercury lamps is caused by the 120 on-and-off arc-strikes when mercury lamps are used on 60-cycle alternating current. The use of two lamps on lead-lag transformer, three lamps on separate phases of a 3-phase supply or the use of incandescent lamps in combination with the mercury type minimizes this effect.

Typical mercury lamp applications:

High Bay Industrial Lighting — Where high level, efficient light output is required, and color rendition is not important.

Floodlighting and Street Lighting— For high level, efficient output from a small source.

Projection Systems — Application limited by color characteristics.

Photochemical Applications—Where ultraviolet output is useful; i.e., chlorination, water sterilization, photocopying.

Black Light—For a wide range of inspection techniques by ultraviolet activation of fluorescent and phosphorescent dyes and pigments.

Sun Lamps — To utilize spectrum lines in the erythemal region of ultraviolet energy, producing suntan.

OVER THE PAST few years, developments in the construction and operation of mercury-vapor lamps have increased their usefulness and boosted their application for all types of industrial lighting, floodlighting and street lighting. With three times the efficiency of incandescent lamps (more light output for given electrical wattage rating), with operating life from about 5 to over 10 times the life of incandescent lamps and with smaller size than fluorescent lamps, mercury-vapor lamps will continue to find wide application.

Mercury-vapor lamps range in size from 100 watts to 3000 watts and are made in a wide variety of bulb shapes. In the 100-watt size, used primarily for special applications and for street lighting, there are PAR spot and flood lamps, tubular lamps,

PS lamps and reflector lamps for flood-lighting and spot lighting. Lamps rated for 175 watts or 250 watts are either BT bulbs or reflector lamps.

Probably the most common mercury-vapor lamp, the one used for many industrial lighting applications, is the 400-watt size. This lamp is available with BT bulbs, tubular bulbs or reflector bulbs. And in the reflector type there is the standard inside reflector lamp—which controls light output in a flood or spot pattern —and the semi-reflector lamp. The semi-reflector lamp has an inside phosphor coating to reflect most of the light out through the flattened face of the bulb but permits about one-third of the light to be transmitted upward through the coating for diffuse uplighting.

Mercury lamps rated for 700 and 1000 watts are available in BT envelopes and in reflector bulbs. These are used where very large amounts of lighting are needed from single sources for industrial and street lighting applications. The 300-watt mercury lamp is a clear glass tube, about 1¼-in in diameter and 55-in long. It is used for high output applications.

ASA Designations

Catalog ordering designations for mercury-vapor lamps now follow the new American Standards Association code for standardized classification of such lamps. Prior to adoption of this code, manufacturers used their own letter and number designations for lamps. This produced much confusion due to the wide variety of non-standard designations. With this new code, developed by the manufacturers in conjunction with the ASA, standard designations of specific lamp types will be the same for all manufacturers.

The ASA designation consists of the letter H (standing for "Hydrargygum," the Greek word for mercury), followed by a number and two letters. The number identifies the electrical characteristics of the lamp and the

REFLECTOR and

semi-reflector lamps (such as those using the R-52, R-57 and BT-56 bulbs) are available with an

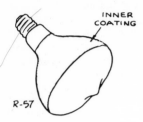

INNER COATING

R-57

internal silver reflector to provide pronounced directional control, or with a phosphor coating acting as a semi-reflector and as a diffuser which directs approximately 1/3 of the light output upward and 2/3 downward.

MERCURY LAMP OPERATION

When current is turned on, mercury is in a liquid state. Initial electric field created between starting electrode and upper operating electrode ionizes the argon gas, and a blue argon arc is established between the two operating electrodes. As the mercury vaporizes, a blue-green mercury arc becomes established. Collisions between electrons and atoms release energy, producing visible light.

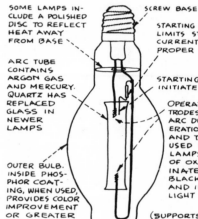

SOME LAMPS INCLUDE A POLISHED DISC TO REFLECT HEAT AWAY FROM BASE

ARC TUBE CONTAINS ARGON GAS AND MERCURY. QUARTZ HAS REPLACED GLASS IN NEWER LAMPS

OUTER BULB. INSIDE PHOSPHOR COATING, WHEN USED, PROVIDES COLOR IMPROVEMENT OR GREATER LIGHT OUTPUT

SCREW BASE

STARTING RESISTOR LIMITS STARTING CURRENT TO PROPER VALUE

STARTING ELECTRODE INITIATES ARC

OPERATING ELECTRODES MAINTAIN ARC DURING OPERATION. TUNGSTEN AND THORIUM USED IN NEWER LAMPS IN PLACE OF OXIDES ELIMINATE ARC TUBE BLACKENING AND INCREASE LIGHT OUTPUT

(SUPPORTS FOR TUBE AND ELECTRODES ARE NOT SHOWN.)

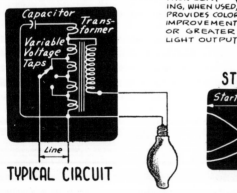

Capacitor

Transformer

Variable Voltage Taps

Line

TYPICAL CIRCUIT

STARTING CHARACTERISTICS

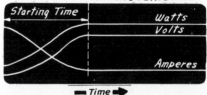

Starting Time

Watts
Volts

Amperes

Time

WHEN ARC IS FIRST ESTABLISHED, current is carried entirely by Argon gas.

AS MERCURY VAPORIZES, it begins to carry part of the current.

AT FULL OPERATING PRESSURE mercury vapor carries the entire current.

type of transformer required. The two letters following the numbers designate the physical characteristics of the lamp. They are arbitrarily chosen letters and are not abbreviations.

The final part of the code designation indicates the color of the lamp light output. A slash mark (/) and the letter "C" indicates color corrected light output for white light for general lighting. The letter "W" indicates white light output with less color correction and higher lumen output. The letter "Y" stands for yellow color for caution or special dramatic effects. A clear lamp with no inside phosphor coating for color correction does not have a slash mark and single letter after the two letters.

The amount and type of phosphors used inside the lamp bulb determine the color character of the light and the efficiency of the lamp. If the clear (uncoated) lamp—including inside frosted bulbs—is taken as 100% of lumen output, it is found that lamps with slight color correction have outputs over 100% and color-corrected and yellow lamps have outputs under 100%—down to 70%.

Longer Life

One of the major improvements in mercury lamps over recent years has been increased life. Improved construction of lamp electrodes uses double-wound coils with new coatings of electron emission material, producing longer life and brighter operation over the life. Cooler operation of electrodes reduces evaporation of the materials. Black deposits of materials on the bulb walls have been eliminated.

Newer lamp types are available with "average economic life" in excess of 12,000 hours—which, in application at 8 to 10 hours per day, comes to 4 or 5 years of lamp life. Because long average life is a major feature of mercury lamps, it is important to know the nature of mercury lamp life.

Throughout the life of a mercury lamp there is a gradual decline in light output due to deposits of emission material on the wall of the arc tube. Some lamps may decline to only 50% of the initial light output. The economics of the cost of light demand that a lamp be replaced before its output has fallen off to this severe extent. It therefore becomes necessary to replace lamps far short of the time at which they would burn out rather than to continue operating them at their uneconomic reduced output. Lighting costs can actually be reduced by replacing lamps at an "economic life" rather than actual life.

Manufacturers' catalogs give rated economic life for various lamps. This life is an estimate of the length of time the lamp should be kept operating in practical installations. Based on lumen maintenance and various power and relamping costs, economic life may vary with different installations. The average economic life of the newer types of lamp is much higher than that of older lamps due to the greatly reduced loss of lumen output over the life of the lamp.

Rated average life of lamps is the average time to burnout of the lamp, and this life value has no relation to lumen maintenance or economics of light production. Lamps with very high maintenance of lumen output have rated economic life very close to or even equal to the rated average life.

Two other figures which reveal the performance capabilities of any mercury lamp are: "initial lumens," which is the light output after 100 hours of operation, and "mean lumens," which is the average light output over the rated economic life.

Other improved characteristics of the newer mercury lamps include: better resistance to thermal shock, to water and snow and to industrial atmospheric gases through use of hardglass, weather-resistant outer bulbs: and dependable starting at temperatures down around 0 deg F.

Fluorescent Lamps

FLUORESCENT lamps operate on the principle of "electric discharge." When proper voltage is applied between the lamp electrodes (one at each end of the tube length), mercury vapor sealed in the tube carries electric current. Interaction between the mercury vapor and the current excites a phosphor coating on the inside of the tube and light is radiated from the outside surface.

Construction

Basically, a fluorescent lamp consists of an inside phosphor-coated length of glass tubing, mercury vapor within the tube, sealed-in electrodes and a base at each end for connection to the electric operating circuit.

Two basic types of electrodes are used in fluorescent lamps: the coated-coil tungsten wire type and the inside-coated metal cylinder type. The tungsten wire type is coated with a material that gives off electrons when heated. This type is used in standard preheat, rapid-start, instant-start and slimline lamps. The metal cylinder type electrode operates at a lower, more even temperature than the tungsten type and is called a "cold cathode." Cold cathode fluorescent lamps use this type of electrode. As a result of electrode operating conditions, cold cathode lamps require higher operating voltages than the other types of fluorescent lamps. Although cold cathode lamps are less efficient, they have much longer life than other fluorescent types.

In addition to low pressure mercury vapor, a small amount of argon gas is sealed in the fluorescent tube to facilitate starting of the arc.

A variety of fluorescent phosphors are used to coat the inside of fluorescent lamps. All of these are excited by ultraviolet radiation from a mercury arc. The combination of phosphors used in a lamp determines the color of light output. Blue, green and pink colored light is obtained by using a color coating on the inside of the tube, in addition to the phosphor coating.

Bases for fluorescent lamps vary depending upon operating characteristics. Preheat-type lamps use bipin bases, one at each end of a lamp. Instant-start hot cathode lamps also have bipin bases, but the two pins at each end are connected together within the base. Slimline lamps and cold cathode lamps have single-pin bases.

Fluorescent lamps are designated as "Type F" lamps. The diameter of a tube is indicated by a "T-number" designation, in which the number tells how many eighths of an inch there are in the diameter; e.g., T-12 indicates a tube diameter of 12/8" or 1½ inches; T-17 indicates a tube diameter of 17/8" or 2½ inches. Standard preheat lamps are available in a wide range of wattages and lengths up to 60 inches with T-5,

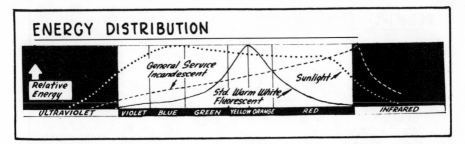

ENERGY DISTRIBUTION

Relative Energy

General Service Incandescent

Std. Warm White Fluorescent

Sunlight

ULTRAVIOLET | VIOLET | BLUE | GREEN | YELLOW ORANGE | RED | INFRARED

T-8, T-12 and T-17 tubes. Slimline lamps range from 42 to 96 inches in length, with T-6, T-8 and T-12 tubes. The 48-inch 40-watt T-12 and the 60-inch 40-watt T-17 are available with bi-pin bases for instant-start operation. The rapid-start lamp is available in the 48-inch 40-watt T-12 size only.

Operation

Average life of a hot cathode fluorescent lamp varies from 2500 to 8000 hours, depending upon the type and size of the lamp and the number of times the lamp is started. Each time a lamp is started, some of the electrode coating is sputtered off. The greater the number of burning hours per start, the longer the life a lamp will have. A lamp will fail when the coating on one of the electrodes is exhausted.

A preheat lamp will blink on and off when one of the electrodes is dead. Such a lamp should be removed promptly to prevent damage to the starter and ballast. Instant-start lamps won't even blink if one of the electrodes is exhausted.

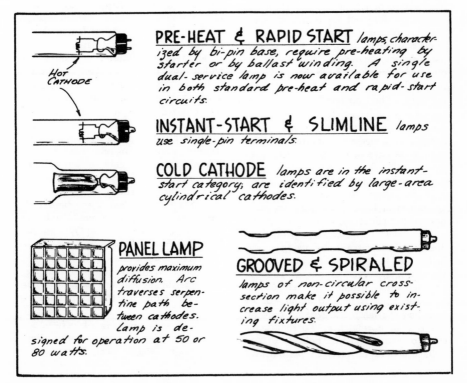

Hot CATHODE

PRE-HEAT & RAPID START lamps, characterized by bi-pin base, require pre-heating by starter or by ballast winding. A single dual-service lamp is now available for use in both standard pre-heat and rapid-start circuits.

INSTANT-START & SLIMLINE lamps use single-pin terminals.

COLD CATHODE lamps are in the instant-start category; are identified by large-area cylindrical cathodes.

PANEL LAMP provides maximum diffusion. Arc traverses serpentine path between cathodes. Lamp is designed for operation at 50 or 80 watts.

GROOVED & SPIRALED lamps of non-circular cross-section make it possible to increase light output using existing fixtures.

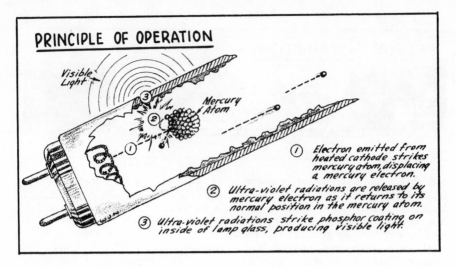

PRINCIPLE OF OPERATION

Visible Light

Mercury Atom

① Electron emitted from heated cathode strikes mercury atom, displacing a mercury electron.

② Ultra-violet radiations are released by mercury electron as it returns to its normal position in the mercury atom.

③ Ultra-violet radiations strike phosphor coating on inside of lamp glass, producing visible light.

Light output of a new fluorescent lamp falls off about 10 per cent during the first 100 hours of burning. Subsequent depreciation in light output is gradual. The light output after the first 100 hours is designated as "initial lumens" output.

Light output is greatly affected by the temperature of the air around the tube. Lamps are designed for optimum operation at ordinary indoor temperatures. Light output decreases with low or high temperatures. Standard lamps can be operated down to 32°F if they are provided with thermal or manual starting switches and operated at higher than normal line voltages. Special lamps and equipment are available for fluorescent lighting at lower temperatures. Instant-start lamps have operating characteristics which make them much better suited to low temperature than preheat lamps.

In fluorescent lighting circuits, voltage higher or lower than rated line voltage reduces efficiency, shortens lamp life.

Fluorescent lamps can be operated on direct current circuits by using a d.c. ballast and a current limiting resistor. Operation, however, is not as good as on alternating current. The light output per watt input is reduced about 40 per cent and the lamp life will be only 80 per cent of normal life. To reduce the effect of end blackening polarity of d.c. voltage should be reversed periodically.

Application

Performance characteristics determine the application advantages of the different types of fluorescent lamps.

Color of light output is an important consideration. Where the effect of natural daylight is required, as in color comparison or color discrimination work, daylight fluorescent lamps should be used. For general area lighting, standard white fluorescent lamps are widely used because of their color similarity to general service incandescent lamps. Soft white lamps are pinker than the standard white and find application where warmer lighting tones are desired. High efficiency standard cool and standard warm lamps offer a wide range of applications for cool or warm lighting tones. For excellent color rendition, at a sacrifice in efficiency, deluxe cool and deluxe warm lamps are best. Colored lamps—blue, green, gold, pink and red—are available for display lighting.

Although all fluorescent lamps are relatively low brightness light sources, optimum conditions of visual comfort and interior appearance require some type of shielding of the lamp.

For viewing moving objects under fluorescent lighting, stroboscopic effect (on-and-off flicker which creates a multiple-image appearance of moving objects) can be reduced by using two lamps on a lead-lag ballast or three lamps on separate phases of a 3-phase system.

Fluorescent lamps are relatively cool and very efficient sources of light, producing better than twice as much light as incandescent lamps of same wattage.

D EVELOPMENT of fluorescent light sources has moved along at a continually accelerating pace over recent years. Improvements in construction and use of new materials has boosted efficiency, produced a wider variety of white colors of output, increased life and expanded application. Some of this development is as follows:

Higher Efficiency—Today's fluorescent lamps take advantage of glass tube shape and new cathode construction to produce more lumens of output for the same watts of input to the lamps. For instance, cooler operating cathodes in 40-watt rapid-start lamps reduce the wattage loss in the cathodes and delivers this wattage to the lamp arc, increasing light output substantially. Such lamps can be used as direct replacements for the older types of lamp. And the use of glass lamp tubes with grooves along both sides of the tube or with the tube made into a twist or spiral has increased the area of the phosphor surface which can be excited to give off greater light. And with such constructions, light output is also increased by the closer proximity of the arc stream to the phosphor surface and by the increased length of the arc because it cannot travel in a straight line from one end to the other.

In the popular 40-watt preheat-rapid start lamp, the more efficient lamps offer 25% and even more light, without increase in cost and for direct replacement in existing fixtures. Such lamps offer simple boosting of light levels in many existing installations.

A new universal lamp is also available which will operate equally well in either preheat or rapid start fixtures, although use of a starter introduces some time delay.

Better Color—Basically, there are seven different "white" fluorescent lamps available to meet the varying requirements for color rendition vs. efficiency:

Cool White—Creates a cool atmosphere for stores, offices, classrooms and factories.

Deluxe Cool White—Contains more red color for better appearance of

---

BUILT-IN COATING in
reflector lamps channels output through small area.

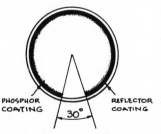

PHOSPHOR COATING

REFLECTOR COATING

30°

Controlled beam through aperture is further concentrated by use of suitable luminaire or reflector.

FLUORESCENT LAMPS are available
in wide variety of types, sizes and ratings.

TYPE	LENGTH	WATTS	DIAMETER
STANDARD PRE-HEAT	6" TO 60"	4 TO 100	5/8" TO 2 1/8"
RAPID START	48" TO 96"	30 TO 100	1 1/2"
INSTANT-START SLIMLINE	24" TO 96"	17.5 TO 74	3/4" TO 2 1/8"
HIGH OUTPUT, HIGH EFFICIENCY	48" TO 96"	30 TO 200	1 1/2"
NON-CIRCULAR CROSS SECTION	48" TO 96"	40 TO 215	1 1/2" TO 2 1/8"
CIRCLINE	OUTSIDE DIAMETER 8 1/4" TO 16"	22 TO 40	1 1/8" TO 1 1/4"

people and merchandise. A general-purpose lamp for color rendition.

Warm White—Creates warm, friendly tone and is especially suited to use with incandescent lamps.

Deluxe Warm White—More red than Warm White, for more flattering appearance of people in commercial areas—lobbies, reception rooms—and at home.

White—For general lighting applications in stores, offices, schools and industrial places. Good to produce high lighting levels, with efficiency, where cool or warm look is not important.

Daylight—Produces blue color of North light of actual daylight, for commercial and industrial work areas. Creates cool atmosphere in industrial plants, and is used to enhance display of white goods.

Soft White—Used in bakery shops and meat display cases, soft peach-pink tint brings out all shades of pink and tan.

Lamp Types

Panel Fluorescent—Although developed some time ago, the panel fluorescent lamp is not yet widely known. The lamp has the shape of a thin panel, about 12 inches square and 1½ inches deep. The face of the lamp has a waffle-like configuration made up of 1¾-in squares, concealing the back-and-forth arc tube behind the waffle. This construction suits the lamp to use with egg-crate louvers which match the waffle sections. The back of the lamp shows the back-and-forth layout of a 5-foot long mercury arc tube which forms a square.

The lamp is designed on a module basis. It is an "area" light source—as compared with regular fluorescent lamps which are "line" sources and incandescent lamps which are "point" sources. The panel fluorescent lamp can be used singly or in groups, assembled as a recessed, surface or pendant luminaire.

Bi-pin terminals are used on this lamp which operates on the rapid-start principle. The terminals are positioned behind an outer rim at adjacent corners—completely within the depth and other dimensions of the unit. Push-pull connectors can be used with the terminals to provide ready insertion with one hand.

Light distribution from the panel is made equally from the top and bottom of the panel. A reflective coating on the top (or back) can be used to direct most of the light out of the bottom (the waffle face) of the panel. Operating at 50 watts, the panel lamp produces 2900 lumens. At 80 watts, it produces 4800 lumens. Rated average life of the lamp is 7500 hours, and the lamp is made in cool white color.

Cold Weather Lamps—The optimum operating temperatures for standard fluorescent lamps is about 77°F. When such lamps are exposed to low tempertures, the glass lamp tubes are cooled and light output drops—by as much as 90% and even more. To meet the need for operating fluorescent lamps at low temperatures, new lamps have been developed over recent years.

One cold-weather lamp is the T-10 1500 ma lamp, for use in enclosed luminaires. The other lamp is the T-10 J 1500-ma lamp equipped with an enclosing glass jacket for use in open luminaires. They are available in 48-, 72- and 96-in lengths. These outdoor type lamps are at maximum efficiency in temperatures of —20°F to O°. They produce about 80% of light output at 40°F. The lamps are used for outdoor installations: bill-boards, display signs, service stations, building floodlighting, shopping centers and parking lots. They are also used for such indoor applications as cold storage warehouses and refrigeration areas.

Aperture Lamp—These are fluorescent lamps with internal reflectors and narrow apertures along their lengths to concentrate the light output along the line represented by the aperture.

The regular fluorescent lamp is really a diffuse line source of light which does not lend itself to precise control of the light output. The aperture lamp is truly a line source and with the proper luminaire provides a beam of light with punch through the aperture.

A typical application for the aperture lamp is in the floodlighting of a building from a cornice or coping on the face of the building. A continuous row aperture luminaire, extending out about 12 inches overall, will afford accurate control of the intense beam from the aperture. Other applications include: wall washing (inside and outside), sign and mural lighting and other cases where a high intensity light output with a sharp cutoff is vital.

Fluorescent Accessories

OPERATION of any fluorescent lamp requires selection and proper application of a suitable ballast. In addition, preheat type hot cathode lamps need some type of starter to facilitate initiation of the arc in the tube. Each of these accessory devices is available in a wide range of capacities and characteristics to suit the requirements of different lamp types and different circuits.

Ballasts

Basically, a ballast is an electrical circuit component used with fluorescent lamps to provide the necessary voltage for striking the mercury arc and then to limit the amount of current flowing through the lamp. Each type of fluorescent lamp requires a ballast designed for its particular operating characteristics. Ballasts differ mainly in the open-circuit voltage they produce for the lamp. Ballasts for use with preheat type lamps produce starting voltages up to about 200 volts. Ballasts for instant-start and multiple cold cathode lamps produce starting voltages from about 450 to 750.

• **Preheat Type Ballasts**—Ballasts for use with standard preheat hot cathode lamps are either simple chokes or chokes and autotransformers. (A "choke" is a coil of wire, wound on a laminated iron core and having the effect of limiting current flow through it.) The fluorescent lamp, because it contains a gaseous discharge, represents what is called a "negative resistance," that is, the current would continuously increase until the lamp was destroyed if some means were not used to limit it. Every ballast must provide sufficient choking action to stabilize the current.

Preheat ballasts for smaller lamps are simply chokes—they limit the current and divide the line voltage (110-125 volts) with the lamp. Ballasts for the 30- and 40-watt sizes of preheat lamps also include an autotransformer to step up the line voltage. The higher voltage is then divided between the lamp and the choke, with each getting about half of the total.

In a two-lamp ballast, a seperate choke coil is used for each lamp. One of the chokes has a capacitor in series with it and the lamp, causing about 120 degrees of phase displacement between currents in the two legs of the ballast circuit. This improves power factor and has the advantage of minimizing stroboscopic (flickering) effect. Of the two lamp-circuits from the ballast, the one with the capacitor in it is called the "lead" circuit; the other, the "lag" circuit. The lamps on these circuits are designated "lead lamp" and "lag lamp."

Two-lamp ballasts for standard 15- to 40-watt lamps require a "compensator" in the starter circuit of the lead lamp. This compensator is a choke coil which provides proper current for pre heating the cathodes of the lead lamp. It facilitates starting and assures normal life and satisfactory lumen maintenance

BASIC CIRCUIT OPERATION

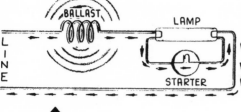

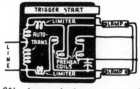

- Starters provide preheat current.
- Compensator aids starting of "lead" lamp.

1 Line current flows through ballast winding, lamp filaments, and starting contacts, heating filaments and setting up magnetic field around ballast.

2 When starter contacts open, magnetic field collapses, generating voltage high enough to establish arc through lamp. Ballast then serves to limit lamp current.

- After lamps start, arrangement of preheat coils bucks out preheat current.

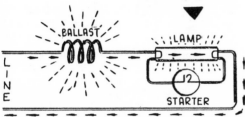

- Lamps operate in series.
- Auto-transformer acts as limiter.
- Preheat current flows continuously.

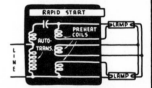

MODIFICATIONS of this basic circuit, shown at right, eliminate first the starting switch, then the necessity for filament preheating.

- High initial voltage starts lamps by "brute force" without preheat current.

STARTERS

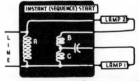

- Voltage across winding C starts Lamp 1; then voltage across A + B + C starts Lamp 2.

GLOW SWITCH

- Current causes glow discharge through gas, closing bimetal contacts and preheating filaments.
- Bimetal cools, opening contacts.
- Inductive kick of ballast starts lamp.

MANUAL NO-BLINK

- Lamp starts normally as with glow switch.
- If not, R heats up, opening C and breaking circuit.
- Pressing button B resets circuit after lamp is replaced.

WATCH DOG

- Lamp starts normally as with glow switch.
- If not, R_1 heats up, opening C and breaking circuit.
- R_2 heats up and holds C open until lamp is replaced.

of the lamp. The starting compensator is built-in in most multi-lamp ballasts of the preheat type. When such a compensator is not part of the ballast, it should be externally added for multi-lamp preheat operation.

Ballasts containing two or three simple series choke coils in parallel are available for use with two or three lamps of the 14-, 15-, 20- or 25-watt size. These units are designed for use in multiple-lamp fixtures.

Many three-lamp industrial fixtures using 40-watt lamps have three-lamp ballasts. These ballasts contain one leading circuit in parallel with two lagging circuits and have power factor of about 90 per cent.

A four-lamp ballast is made to operate two 85-watt lamps in series on each leg of a modified two-lamp ballast. Application of this arrangement is restricted to supply circuit voltages in the ranges of 199-216, 220-250 or 250-280 volts.

Trigger-start ballasts are available for split-second starting of preheat lamps, without using starters. This is accomplished by cathode-heater windings in the ballast. These ballasts are made for single-lamp operation of 14-, 15-, 20- and 32-watt lamps. With this type of operation, lamp life is reduced about 20 per cent.

Rapid-start ballasts are available for use with two 4-foot 40-watt rapid-start lamps. Such a circuit uses no starters; special cathodes in the lamps are heated directly by the ballast. Preheat or instant-start lamps cannot be used on rapid-start ballasts.

• **Instant-Start Ballasts**—Ballasts designed for use with slimline and instant-start lamps have only two functions: provide proper starting voltage and limit current. In these circuits, there are no starters or cathode preheating. Generally, because these ballasts must provide considerably higher voltages, they are physically larger and have somewhat greater electrical losses than preheat type ballasts. Slimline ballasts are available for lamp currents of 120, 200, 300 and 430 milliamperes.

Instant-start ballasts are made for mul-tiple- and series-type operation. Series-type ballasts are made for instant-start and slimline lamps.

Hot or cold cathode fluorescent lamps may be operated on series-type or multiple-type circuits. Special lamps are recommended, however, for series circuits. Cold cathode fluorescent lamps are commonly operated on high voltage series circuits. Transformers for such circuits are rated in milliamperes—the maximum current-carrying capacity of the transformer depending upon the lamp used. Sizes of these transformers range from 2,000 to 15,000 volts, 18 to 120 ma. (milliamperes).

Resistance ballasts (no choke coil) are used to limit current flow through a lamp on a direct current circuit. An inductance coil is used, however, to give a voltage "kick" to start the lamp.

• **Ballast Construction**—A typical ballast is enclosed in a metal container filled with a heavy impregnating compound which surrounds coils and capacitor. Construction is such as to radiate heat and minimize hum. Ballasts create heat, however, and should be ventilated.

Starters

Starters are used only with preheat type lamps and ballasts. The purpose of the starter is to complete a circuit for current flow through the lamp filaments, and then quickly break the circuit when the filaments are sufficiently heated. The basic type of starter is an ordinary push-button, in series with the lamp filaments across the supply voltage. By depressing the button, current flows through the filaments. After a few seconds, the button is released, and a voltage "kick" from the choke jumps the mercury arc between the heated, electron-emissive cathodes.

Several types of starters are available for accomplishing the starting operation automatically upon closing a line switch. The glow-switch, no-blink, watch dog and thermal switch starters are specific types. Basically they all do the same job. The no-blink and watch dog types prevent on-and-off blinking due to repeated attempts to start dead lamps.

Luminaires

PRIMARILY, every lighting fixture is a control device. A luminaire, the proper word to describe what we usually call a lighting fixture, must take "bulk" light as produced by a bare lamp and control its use in making vision possible. An understanding of light characteristics and the fundamentals of light control is essential to effective application of luminaires.

Light Control

The purpose of all lighting is to make vision possible. As a result, the quality of any lighting application is directly proportional to the degree to which the lighting satisfies the requirements of the human eye. These requirements are indicated by four factors in vision:

1. **Size**—The closer, larger object is more easily seen than the distant or smaller object.

2. **Brightness**—The higher the light intensity on an object, the greater its brightness and the more easily it is seen.

3. **Contrast**—The greater the contrast in brightness or color between an object and its background, the easier it is to see the object.

4. **Time**—Seeing requires time. The higher the lighting level on an object, the less time required to accurately "see" the object; and vice versa.

Of these factors, only brightness and contrast are related to the control job which a luminaire must do. These factors, however, can minimize the effects of small size and limited seeing time. All of the factors are related to luminaire selection.

Light control as provided by different types of luminaires is much like the control of water spray due to different settings of the nozzle on a garden hose. Just as the water from the hose may be sprayed in a hard blast or a widespread mist, the "bulk" lumens from a lamp may be focused in a high intensity beam by one luminaire or diffused in many directions by another. The essential idea is that the luminaire will give some form of direction to the travel of light.

Although specific control tasks vary with different types of luminaires, there are certain characteristics which almost all luminaires provide:

• A luminaire should "spray" a maximum amount of lumens on the immediate visual task. The more lumens on the work, the higher the foot-candle level, which is just lumens-per-square-foot.

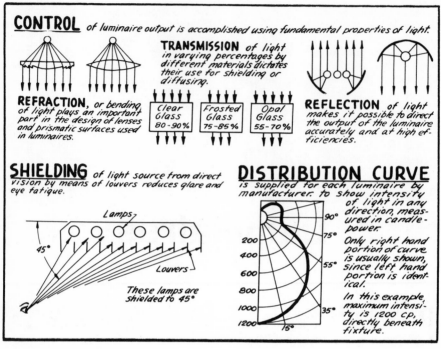

CONTROL *of luminaire output is accomplished using fundamental properties of light.*

REFRACTION, *or bending of light plays an important part in the design of lenses and prismatic surfaces used in luminaires.*

TRANSMISSION *of light in varying percentages by different materials dictates their use for shielding or diffusing.*

Clear Glass 80-90%
Frosted Glass 75-85%
Opal Glass 55-70%

REFLECTION *of light makes it possible to direct the output of the luminaire accurately and at high efficiencies.*

SHIELDING *of light source from direct vision by means of louvers reduces glare and eye fatigue.*

Lamps
45°
Louvers
These lamps are shielded to 45°

DISTRIBUTION CURVE *is supplied for each luminaire by manufacturer, to show intensity of light in any direction, measured in candle-power.*

Only right hand portion of curve is usually shown, since left hand portion is identical.

In this example, maximum intensity is 1200 cp, directly beneath fixture.

90°
75°
55°
35°
15°
200
400
600
800
1000
1200

• A luminaire should keep lumens (glare) out of the eyes. Glare in the eyes cripples vision.

• A luminaire should provide a controlled diffusion of light.

How luminaires effect control of light output is, of course, a function of their construction. A luminaire for incandescent or gaseous discharge lamps may have one or more aluminum, porcelain or baked white enamel interior reflecting surfaces which redirect various portions of the lamp lumen output. Louvers or baffles may be used to diffuse the light, shield the bare lamp and eliminate objectionable surface brightness of the unit. A wide range of diffusing glass and plastic and lighting control lenses are in common use for shielding and control.

Luminaire Selection

Selection of the best type of luminaire for a particular lighting installation is part of the over-all design and depends upon all of the factors in the design. Whether the light source to be used is incandescent, fluorescent or mercury vapor; whether the area to be lighted is interior, exterior, commercial, industrial or residential; what the size and shape of the area are; what the economy of the situation happens to be—these are all considerations which have been made and resolved before selection of the luminaire.

Typical design factors, as they relate to luminaire selection, are:

• **Light Distribution** — Uniformity of lighting level throughout an area is important in many installations where visual tasks are continuous and rigorous. In such cases, the width of light spread by various luminaires must be related to the number of units to be used, the mounting height and the spacing between units. Of course, where design calls for accent lighting of certain local areas and specific light coverage in a vertical plane, the problem of luminaire selection is quickly narrowed to those units which have the required distribution pattern.

• **Light Quality**—In any installation, the required quality of lighting will usually limit the number of luminaire types from which selection must be made. To minimize direct glare, a flush-mounted unit may be preferred over a stem-mounted unit because the extra height of the luminaire places it above the normal viewing angle. Again, a luminaire with luminous side panels may be called for to reduce brightness contrast between the unit and its surroundings. The size and shape of the interior to be lighted greatly affects the type of luminaire required for a particular degree of visual comfort.

• **Light Diffusion**—The extent to which diffusion of the light output is desirable is still another clue to the most suitable luminaire. For critical seeing tasks, in offices and schools and in areas where highly reflecting surfaces abound, maximum diffusion of light is desired. Such diffusion is produced by widely spacing a large number of small area luminaires, by using large area luminaires or by some type of indirect lighting.

Luminaire Construction

Luminaires, for both incandescent and gaseous discharge lamps, are available in a wide range of types, sizes and appearances. Luminaire characteristics which should always be checked are:

• **Efficiency**—A percentage of the lamp lumen output which the luminaire "puts to work." Generally, the greater the amount of shielding and light control available from a luminaire, the lower is the efficiency. This is due to absorption and loss of light. Of course, many installations absolutely require specific lighting characteristics which justify the lower efficiency due to control.

• **Physical construction** of a luminaire should be sturdy and compact—for the type of service for which it is designed. Ease of maintenance—cleaning, relamping, replacement of parts—should be considered.

• **Electrical quality** of a luminaire is usually attested to by the Underwriters' label. First-class wiring, wiring devices, sockets, boxes, splices, auxiliary electrical elements, etc., are important to efficient, trouble-free operation.

• **Appearance** of a luminaire should be suited to the decorative and architectural themes of an interior.

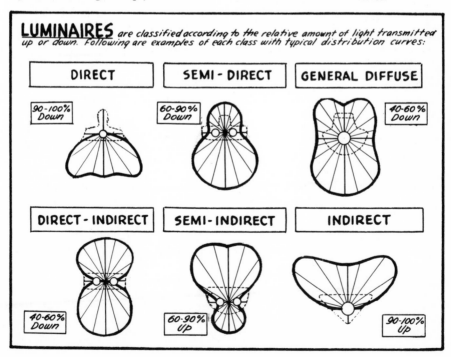

LUMINAIRES are classified according to the relative amount of light transmitted up or down. Following are examples of each class with typical distribution curves:

| DIRECT | SEMI-DIRECT | GENERAL DIFFUSE |
| 90-100% Down | 60-90% Down | 40-60% Down |

| DIRECT-INDIRECT | SEMI-INDIRECT | INDIRECT |
| 40-60% Down | 60-90% Up | 90-100% Up |

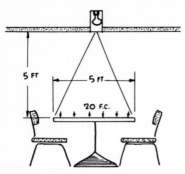

LOW-VOLTAGE LIGHTING *units*
(recessed, surface-mounted, adjustable) are now available which can economically fulfill a variety of residential and commercial lighting requirements for both modernization and new work.

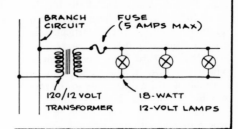

REFLECTOR LAMPS *in units designed for 12 volts can provide intimate lighting on tables such as in cocktail lounges.*

LUMINAIRES are available in an extremely wide range of types and sizes to provide maximum effectiveness in the use of incandescent, fluorescent and mercury vapor lamps in commercial, industrial, institutional and residential buildings and outdoor applications.

Incandescent Luminaires

• **To meet the trend to lower ceiling heights and the use of suspended ceilings in modern buildings,** a broad range of recessed incandescent luminaires is available for various types of lamps and ceiling constructions.

In addition to standard 120-volt luminaires for recessed mounting, there are recessed units available which make use of the 12-volt, R-12, 18-watt reflector lamp with a bayonet base. Such units offer effective decorative and functional lighting in areas with low ceilings or where only a low

lighting intensity is required. In restaurants, taverns, offices, churches and many other areas, these inconspicuous 12-volt units can be used alone or in combination with larger wattage 120-volt units.

The 12-volt units can be laid out with close or wide spacing, depending upon the light level required and the architectural needs of the interior. In restaurants or cocktail lounges, a single 12-volt, 18-watt unit mounted about 5 feet above the center of each table will provide 20 foot-candles in a spot 5 feet in diameter at the table top.

A very big advantage of the 12-volt lighting units is National Electrical Code permission to wire the units in accordance with Class 2 signal wiring from Article 725 of the Code. This means that boxes do not have to be used at outlets and lamp cord or other signal wire may be used to wire the luminaires from the 12-volt secondary of the step-down transformer supplied as part of the lighting system. Conduit, EMT, armored cable or other conventional 600-volt wiring methods are not required, provided the transformer supplying the lighting units is not rated over 100 va (about 100 watts, so that it can take 5 of the 12-volt, 18-watt units) and the circuits to the luminaires are protected by fuses rated not over 5 amps. This use of lamp cord can be a terrific economy factor in modernizing of lighting in commercial and residential interiors.

• **Growing application of quartz-iodine lamps** has brought development of new and more versatile luminaires for the 500-watt and 1500-watt lamp tubes. These are weatherproof units designed to hold the lamp tubes in their required horizontal mounting position, and finned construction of the cast metal housings is designed to dissipate the heavy heat load of these high intensity lamps. These units are made with reflectors which take advantage of the small, pencil-like shape of the tube to provide rectangular pattern of the light output, using the light output much more efficiently than reflectors with standard-shape incandescent bulbs. The small diameter of the lamp tube minimizes the trapping of light behind the lamp tube. Thus, the reflector utilizes just about all of the lumens from the lamp.

Quartz-iodine luminaires are made with reflectors for three general beam spread patterns: narrow, medium and wide vertical beam spread. Some units have reflector-lamp assemblies which can be adjusted continuously, by means of an external handle, for beam spread anywhere between narrow and wide. And position-adjusting knuckle joints are provided with calibrated marks for various angles of the housing with relation to the mounting.

Fluorescent Luminaires

• **Although surface and suspended luminaires find wide use for fluorescent lighting in industrial and commercial occupancies,** especially schools, recessed fluorescent luminaires have become the general type for use in new construction for general area lighting in stores and office buildings and other interiors using suspended ceilings.

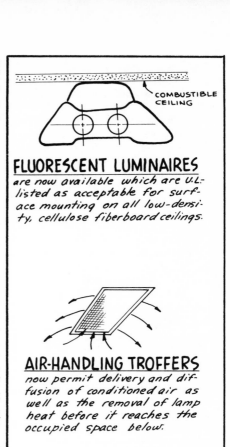

COMBUSTIBLE CEILING

FLUORESCENT LUMINAIRES are now available which are U.L.-listed as acceptable for surface mounting on all low-density, cellulose fiberboard ceilings.

AIR-HANDLING TROFFERS now permit delivery and diffusion of conditioned air as well as the removal of lamp heat before it reaches the occupied space below.

BEAM CONTROL

QUARTZ-IODINE LAMP

FLOODLIGHTS using quartz-iodine lamp permit control of high-intensity light beam pattern from narrow (left) through full beam (right).

• **The most recent form of the recessed fluorescent luminaire finding expanding application is the air-handling troffer—a combination lighting unit and air-conditioning diffuser.** Since air-conditioning is now standard in commercial buildings and the module layout is generally used, the air-handling troffer provides two functions from a basic unit which can be readily fitted into the module. Such luminaires may serve as either air-supply or air-return units. In a given layout, half of the luminaires, say, are used for supply and the rest for return. Or all of the units may be used for supply with some other means of air return. Each air-handling troffer includes an adjustable air valve to effect proper control of the air flow through it, as required by the design of the particular air-conditioning system.

Another phase of fluorescent luminaire development is that for outdoor "area lighting" or floodlighting. Street and highway luminaires now make use of large area fluorescent-lighted enclosures of weatherproof construction, listed by UL and provided with rust-resistant hardware and latches. Such fluorescent floodlights are designed to be mounted like wings on the top of poles and standards for lighting parking lots, gas stations and outdoor exhibits. Layouts of such standards can provide intensities from 2 to 5 footcandles.

Mercury Vapor Luminaires

• **Continued growth in the application of mercury-vapor lighting for industrial high bay lighting and for lighting of airplane hangars and similar buildings has pushed development of better luminaires for general area mercury lighting.** Large reflector housings, round or square in cross-section, are available with or without diffusing or control lens covers. Units are available with special couplings between the ballast assembly and the lamp housing, so that the only installation wiring to the luminaire is that made to the ballast unit. The lamp housing is readily attached or detached by a simple, tight connector which makes the electrical connections between the ballast and the lamp. Other luminaires are available for remote mounting of the ballasts.

Still other new luminaires have been developed for use of mercury-vapor lamps in floodlighting of sports arenas and stadiums, building façades, parking areas, etc. Some luminaires are the round cross-section type used also with incandescent lamps. Others have rectangular cross section and provide a rectangular pattern of light output to match the generally rectangular shape of areas being lighted.

Electric
Space Heating

ELECTRICITY represents a versatile source of heat for interior spaces used for human occupancy. It can be applied to space heating in a wide range of methods; it is simple, reliable and easily controlled in the production of thermal energy. As a result, electric space heating equipment is available in many different forms, each with its own installation and application advantages.

Operation

When electric current flows through electrically-conductive material, the movement of electrons within the material causes friction and heat is given off as a result. Although this action takes place in all electric circuit wiring, the heating effect is minimized by the low resistance of circuit wires. In electric space heaters, the heating element is made in such a way as to maximize and concentrate the heating effect of current.

A heating element is some form of electric resistance unit—wire, plate, strip, tube, panel. Heating element resistors are usually made of metal alloys.

Nickel-chromium (nichrome) wire or ribbon and non-metallic compounds containing carbon formed into rods or other shapes are used in typical heating elements. Resistors in heating elements may be in the form of exposed coils mounted on insulators or of metallic conductors embedded in refractory insulating material and protected by a metallic sheath.

Strip elements are found in some convection air heaters and low-temperature radiant heaters. Ring and plate elements are used in some small air heaters. Metal or oxide conductive films on glass and ceramics are used in electric heating panels. Tubular elements are used in air heaters. Some radiant panel heaters have heating elements that consist of resistor wire and asbestos or glass thread woven into a fabric. In applications where radiant heat alone is desired, infrared lamps are used as heating elements, with tungsten filaments as resistors.

All of the electric energy applied to a resistor is transformed into heat. The wattage rating of a heating element is, therefore, equal to the heat output in watts.

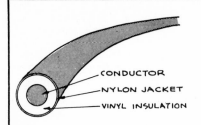

HEATING CABLE *for use in ceilings and floors has a conducting element designed to produce about 2 3/4 watts per foot of length when its two ends are connected across rated voltage.*

O VER RECENT years, rapid growth in popularity of electric space heating has stimulated development of electric heating equipment, clarified application advantages and produced a continuing flow of new types of equipment into the market. Because operation, installation and performance of the different types of equipment vary widely, effective application of electric space heating demands a clear and comprehensive understanding of all of the various devices available for use.

Wall Units

Unit electric heaters for mounting in or on walls are either radiant-convector units or fan-forced-convector units. The radiant-convector units have large-area radiating surfaces—varying from rectangular—and square-shaped wall panels to the long, narrow shape of baseboard units. In such units, heat from the electric element is transferred to a metal or glass radiating surface which sends the heat into the room. Although these panels are commonly called "radiant" panels, only about 50 or 60% of the heat output from them is delivered to the room

as radiant energy. The balance of the heat output is delivered to the room by convection air currents which circulate across the panels, pick up heat and then rise in a convection pattern.

Furniture should not be placed directly in front of wall panels because convention will be impaired and radiated energy will be absorbed by the furniture.

Baseboard heating units have enjoyed wide acceptance because of their ability to deliver plenty of heat from a long, relatively low temperature source which, as a result of dimensions (4 to 9 inches high, by 2 to 3 inches deep), is unobtrusive and structurally integrated in a room, while practically eliminating concern over heat blocking due to furniture placement. Baseboards are made in lengths of 1 to 10 feet, from 100 to 3000 watts, with a range of accessories—corner sections, blank sections, thermostats, other controls, etc.

Although most baseboard units have the heat directly transferred from the heating element to the radiating fins and surfaces of the unit, there are units which use water as the transfer medium. One such unit contains its own water supply and heating ele-

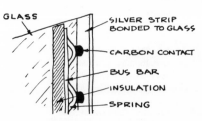

GLASS PANEL *heater is shown from the rear with back housing removed. Current is introduced to conductive coating on glass through bus bars and contacts on opposite edges of glass.*

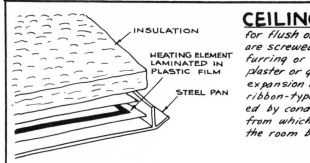

CEILING PANELS

for flush or recessed mounting are screwed to ceiling joists or furring or may be fastened to plaster or gypsum board using expansion anchors. Heat from ribbon-type element is imparted by conduction to steel pan, from which it is radiated to the room below.

INSULATION

HEATING ELEMENT LAMINATED IN PLASTIC FILM

STEEL PAN

ment. The element heats the water which circulates in the unit and gives up the heat to radiating fins which transfer the heat to the room by convection and radiation. The water-type units offer very even heat output due to thermal inertia of the water, i.e., the water temperature stays fairly constant while the heating element is going through appreciable temperature variations due to ON-OFF cycling of the thermostat, keeping the temperature of the radiating surfaces of the unit fairly even. Commonly referred to as "hydronic" heating units, such devices are available in sections up to 25 feet, up to 4800 watts.

Wall panel heaters are, compared to baseboard units, small-area and high-temperature heaters. These are shallow, surface mounting units. To do the same heating job as a baseboard unit which has many times the radiating surface of the wall unit, the wall unit must operate much hotter. As an analogy to light sources, an incandescent lamp which puts out as much light as a fluorescent lamp must operate with a greater brightness.

Furniture placement as a heat block is of particular concern when a wall panel is used.

In many wall-mounting unit heaters, the inclusion of a fan or blower permits greater quantities of air to be drawn past the heated fins or element, thereby increasing the convection part of the heat output. Blowers are necessary in larger heater units for commercial and industrial areas to provide controlled directing of a heated air stream to specific spaces to be heated and to provide better temperature uniformity in large areas.

One very new concept in forced convection wall heater units is designed for installation in the wall with only a 5-in. high, flat grille showing on the wall surface. The unit fits the space between standard wall studs and is made for 1000, 1500 or 2000-watt energy input.

Ceiling Techniques

Electrically heating a room from the ceiling is done either by embedded cables or by heating panels. Because electric heat units at the ceiling must direct heat down into the room, such

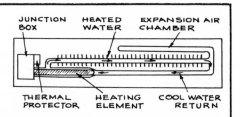

HYDRONIC BASEBOARD

has sealed-in water supply. Element heats water, which circulates constantly by convection; heat is transferred to air from finned tubing.

JUNCTION BOX

HEATED WATER

EXPANSION AIR CHAMBER

THERMAL PROTECTOR

HEATING ELEMENT

COOL WATER RETURN

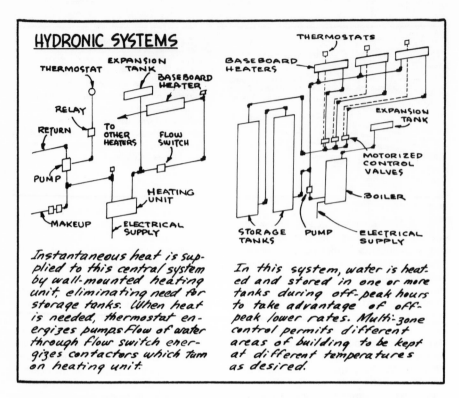

HYDRONIC SYSTEMS

THERMOSTATS

THERMOSTAT

EXPANSION TANK

BASEBOARD HEATER

BASEBOARD HEATERS

RELAY

EXPANSION TANK

RETURN

TO OTHER HEATERS

FLOW SWITCH

MOTORIZED CONTROL VALVES

PUMP

HEATING UNIT

BOILER

MAKEUP

ELECTRICAL SUPPLY

STORAGE TANKS

PUMP

ELECTRICAL SUPPLY

Instantaneous heat is supplied to this central system by wall-mounted heating unit, eliminating need for storage tanks. When heat is needed, thermostat energizes pumps Flow of water through flow switch energizes contactors which turn on heating unit.

In this system, water is heated and stored in one or more tanks during off-peak hours to take advantage of off-peak lower rates. Multi-zone control permits different areas of building to be kept at different temperatures as desired.

units must have a very high percentage of their heat output of the radiant type. Convection heating depends upon the ability of heated air to rise, so convection output from units at the highest part of a room is essentially blocked by the air layer in contact with the heated surface of the ceiling or panel on the ceiling, which is the hottest air in the room. Some convection currents are produced by air contacting the floor of the room and other objects which have been heated by radiation from the ceiling.

Heating from the ceiling offers a number of distinct advantages: **1.** The equipment takes up no livable or usable space. **2.** It does not interfere with wall decoration. **3.** Furniture placement is unrestricted.

The use of cable embedded in the ceiling makes the entire ceiling area a radiating surface. In plastered ceilings, the cable is stapled in a back-and-forth grid pattern on the rough plaster coat, with the cable ends

brought down in a wall to the thermostat. The finish plaster coat is then applied over the cable, completely embedding it. Or, where dry-wall construction is used on the ceiling, the cable is stapled between a sandwich of plasterboard, with a layer of plaster covering the cable between the plasterboards.

Heating cable designed for ceiling (and floor) systems is a plastic-insulated, nylon-jacketed alloy wire rated to dissipate about 2¾ watts per foot of length. Requirements for different heat loads are met by varying the length and spacing of the cable within the ceiling. Each cable is furnished with a lower resistance, non-heating lead spliced to each end, for connection to branch circuit and thermostat.

Steel ceiling panels are shallow panels—2 by 4 ft or 2 by 5 ft—for recessed or surface mounting, with ratings of 500 or 700 watts· These units have their heating elements (such as aluminum ribbon type) sealed in a laminated form with leads brought to a junction box from which non-heating leads can be run in flex to the branch circuit. Similar surface panels are made using glass with electrically conductive coating.

Surface-mounted, glass-panel ceiling heaters, 2 in deep, are made in 18-in and 24-in squares, rated 600 and 1000 watts. A number of units can be combined in multiples of the basic module to achieve various patterns and heat outputs.

Heating cable can be embedded in the floor, instead of the ceiling, of a room. In such installations, the entire floor area becomes the heat radiating source. As in the case of ceiling heat cable, a floor cable installation takes up no livable space within the room, does not interfere with decorations and does not affect furniture placement. The heat output from so large a source is even and well distributed.

Installation of floor cable heating systems is made in concrete slab floors and involves a simple, straightforward procedure. First, the concrete is poured over a vapor barrier, forming a slab of less than full finished depth. Next, the heating cable is fastened to the slab in a back-and-forth grid of parallel runs—on a predetermined spacing to obtain the required watts-per-square-foot of heat density. Because the cables are designed to produce about 2¾ watts per foot of length, length and spacing of the total run determine the total input to any given area, as described for ceiling cables. The final step of the installation consists of pouring about another inch of concrete over the cable to bring the slab to its full depth.

For heated floor installations, heating cable is also available as factory-assembled mats of cable preformed into back-and-forth grids. These mats are made in various widths and lengths to provide about 10 to 15 watts-per-square-foot of heat output, depending upon voltage, for supplementary heat in floor slabs of bathrooms, playrooms, basements, garages, breezeways, etc. Floor tile is usually used where a finished covering on the floor is desired. And to avoid foot discomfort, floor temperatures are normally kept below 85 deg F.

In a heated floor system, the large mass of heated concrete could cause periods of discomfort when rapid or sudden changes of outdoor temperature make it impossible for the mass

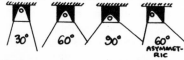

INFRARED HEATERS, directed at work position from at least two directions, provide instantaneous local radiated heat where and when needed without heating the intervening air.

30° 60° 90° 60° ASYMMETRIC

Fixtures are available for one or more elements, lamps, or tubes. Reflectors direct infrared beam in narrow or wide symmetrical or asymmetrical pattern, as required.

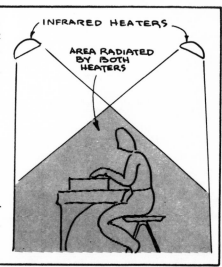

INFRARED HEATERS

AREA RADIATED BY BOTH HEATERS

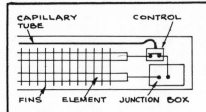

LINEAR CONTROLS *protect baseboard heaters against high temperature buildup caused by obstruction of air flow by blankets, pillows, etc. Liquid-filled capillary tube extends the length of the heater; excessive heat causes liquid to expand, opening contacts in control unit and thus stopping the flow of current to the element. When liquid cools, contacts close, and the control continues to cycle until the obstruction is removed.*

of concrete to increase or decrease its temperature fast enough to maintain constant temperatures in the indoor space. Heating cable in concrete is best suited to spaces which are occupied fairly continuously and held at relatively constant temperature.

Infrared Heaters

More effective radiant type of heating than that produced by the wall and baseboard heating units described so far can only be achieved by operating the heating element at temperatures much higher than those used in the wall and baseboard units. Electric heating units employing quartz lamp or quartz tube elements operate around 4000 deg F and 1800 deg F respectively and produce a much higher percentage of infrared radiation in still air—ranging to 80% or more of the total energy output.

The quartz lamp and quartz tube heaters reach their operating temperatures rather rapidly. The quartz tube glows with a red heat. The visible radiation of the quartz lamp is a whiter light and is frequently used to fulfill both heating and lighting requirements.

Another high-temperature heating source utilizes metal sheath elements operating at temperatures around 1300 deg F. Such devices produce about 60% of their output as infrared heat radiation. These metal elements also operate with a dull red glow.

High-temperature infrared heaters are highly directional, i.e., their effect is similar to a fireplace. They do not heat the air; their radiated energy is absorbed as heat only by persons or objects in the direct path of the radiation. Because of this directional characteristic, infrared heaters used to heat people should be placed so as to radiate at a body from at least two opposite directions.

All three types of infrared heaters are used for applications where heating the intervening air would be impractical or costly. Typical applications for infrared units are: in industrial buildings over machines or other work positions; outdoors in partially sheltered places like loading docks, sports grandstands; outdoor bus or train waiting platforms or shelters; buildings with large air-space volume or high ceilings, such as gymnasiums.

Central Systems

A number of central electric heating systems are available to provide complete heating of all of the rooms in a house or building:

1. The central electric furnace system is much like that used with forced air combustible fuel systems, with the substitution of an electric heating element for the fuel burner. With adequate duct insulation and modulating temperature controls combined with multi-stage element operation, operating cost can be close to that obtained with individual room heaters (although initial installation costs are likely to be higher). The central furnace system simplifies the summer-cooling problem, since the same ductwork may be used.

One complete comfort-conditioning system of the central type includes heating, cooling, air cleaning, fresh air ventilation, humidification and dehumidification. A central unit installed in basement, garage or laundry area contains air filter, humidifier, cooling coil and central heating bank. The central heater tempers air supply which is ducted to individual forced-air baseboard units in each room. An additional heat element in each baseboard unit is controlled by a room thermostat to provide extra heat above the temperature of central air, as needed.

2. Electrically energized hot water heating systems are generally similar to the combustion fuel hot water systems which have been widely used. In one typical system, the central water heating unit—rated at 7.2, 15 or 24 kw—is wall mounted in the basement or a closet and feeds hot water to a loop of piping with baseboard radiator units in series. As the water circulates through the loop it is heated when it passes through the central heating unit; no storage tanks are required. When heat is needed, a thermostat turns on a circulating pump. The movement of water turns on a flow switch, which energizes contactors controlling the heating element. A small tank for domestic hot water may be connected into the same piping.

Totally different is another hot water system which uses a boiler and one or more water storage tanks. Individual baseboard sections are connected in parallel, making zoning possible. Motorized zone control valves control the flow of hot water from the storage tanks to the baseboard radiation in accordance with the action of thermostats located in each zone. This system may take advantage of lower off-peak electrical rates where such rates are made available by the utility. Water is heated during the night and stored for use during the next day.

EFFECTIVE application of eletric space heating equipment depends upon proper use of control devices to regulate the heat output of the heaters.

Although there is a variety of control devices available for the purpose, the majority of heat control needs can be satisfied by single-pole line-voltage and low-voltage thermostats. The line-voltage thermostat acts directly as a line switch in the circuit supply to the heater load, turning ON and OFF the operating current to the load. The low-voltage thermostat acts as a switch in a control circuit, turning ON and OFF the operating current to a relay which, in turn, uses its contacts to control load current to the heating equipment.

Each type of thermostat has its own advantages. The line-voltage thermostat is the simplest, most direct and most economical type of heat control. But the line-voltage unit requires a wiring box and must use a standard power-and-light wiring method — non-metallic or armored cable or conductors in conduit (local codes should be checked). On the other hand, low-voltage control circuits usually offer the advantage of simple "bell" wire interconnections, run without any raceway and stapled in place. The low-voltage thermostat can be easily screwed to the wall surface. And since the low-voltage thermostat is called upon to control only the low-level operating current of a relay, it can be made more sensitive to temperature changes and offer a finer type of control than a line-voltage unit. But, then, low-voltage controls cost more, and additional labor is involved in installing the relay units.

In quiet places, the noise of the snap-action type of contacts in some line-voltage thermostats can be objectionable. And there are some relay units used with low-voltage systems which also produce an audible intrusion on quiet areas. There are, however, certain models of line-voltage

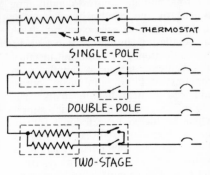

LINE-VOLTAGE THERMOSTATS

in their simplest form are equivalent in function to single-pole switches. Where required to serve also as circuit disconnects, they incorporate an "off" position, and both sides of the line are broken. Modulating control is provided by two-stage thermostat, the two contacts being set ½ to 1½ degrees apart so as to energize first one heater element and then both, depending upon outside temperature.

HEATER

THERMOSTAT

SINGLE-POLE

DOUBLE-POLE

TWO-STAGE

thermostats and low-voltage relay units with inaudible operation.

Although the thermostat controls thus far described are wall-mounted on an inside wall of the area to be heated, line-voltage thermostats are also used built into the heater unit, with a sensing bulb in the heater air stream. This is, perhaps, the least expensive type of room heat control, although there are objections to it. First, the large mass of the sensing bulb is slow to detect changes and hence is slow cycling. And, then, the position of the bulb in the heater does not permit it to properly sample actual room conditions.

Several types of line-voltage thermostats are in use. Single-pole units serve single-stage heating needs. Double-pole units are used to break both conductors to a single-phase heater unit, thereby satisfying the re-

quirements for disconnecting means of NE Code Section 422-46, which says, "Thermostats . . . which indicate an OFF position and which interrupt line current shall open all ungrounded conductors in the OFF position." It further states, "Thermostats . . . which do not have ON or OFF positions are not required to open all ungrounded conductors."

Modulating controls are used with 2-stage heaters to provide two levels of heat output—a low output level, with one stage of the heater operating; and a high output level, with both stages of the heater operating. The controller has two sets of contacts, each one controlling current flow to one of the stages. One set of contacts is set to close at ½ to 1½ degrees below the temperature at which the other set closes. With moderately cold outside temperature, the first set of

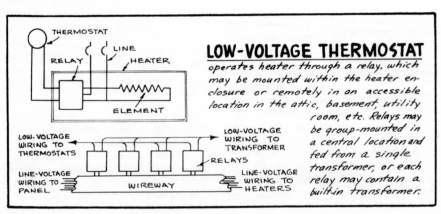

THERMOSTAT

LINE

RELAY

HEATER

ELEMENT

LOW-VOLTAGE WIRING TO THERMOSTATS

LINE-VOLTAGE WIRING TO PANEL

WIREWAY

LOW-VOLTAGE WIRING TO TRANSFORMER

RELAYS

LINE-VOLTAGE WIRING TO HEATERS

LOW-VOLTAGE THERMOSTAT

operates heater through a relay, which may be mounted within the heater enclosure or remotely in an accessible location in the attic, basement, utility room, etc. Relays may be group-mounted in a central location and fed from a single transformer, or each relay may contain a built-in transformer.

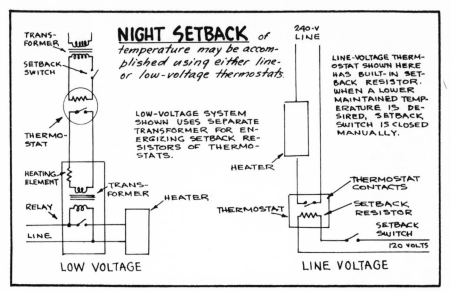

NIGHT SETBACK of temperature may be accomplished using either line- or low-voltage thermostats.

TRANS-FORMER

SETBACK SWITCH

THERMO-STAT

HEATING ELEMENT

RELAY

LINE

TRANS-FORMER

HEATER

LOW-VOLTAGE SYSTEM SHOWN USES SEPARATE TRANSFORMER FOR ENERGIZING SETBACK RESISTORS OF THERMOSTATS.

HEATER

THERMOSTAT

LOW VOLTAGE

240-V LINE

LINE-VOLTAGE THERMOSTAT SHOWN HERE HAS BUILT-IN SETBACK RESISTOR. WHEN A LOWER MAINTAINED TEMPERATURE IS DESIRED, SETBACK SWITCH IS CLOSED MANUALLY.

THERMOSTAT CONTACTS

SETBACK RESISTOR

SETBACK SWITCH

120 VOLTS

LINE VOLTAGE

contacts will close when the inside temperature falls below the thermostat setting. This will energize one stage of the heater which will be sufficient to match the heat loss to the outside and will cycle as necessary to maintain inside temperature at the desired level. If the outside temperature is very cold, producing a higher heat loss from the heated space, the one stage of the heater will not be able to match the heat loss. The inside temperature will, therefore, fall the extra degree or so and the second set of contacts will then energize the second stage of the heater. The two stages of the heater will then cycle as necessary to maintain the desired inside temperature.

Another technique which provides two levels of heat output from a heater to meet the demand of moderate or severe cold weather uses a 2-stage low-voltage thermostat. The thermostat has two sets of contacts with about 1 degree difference in the temperature levels at which they operate, the same as the modulating line-voltage thermostat described before. The thermostat works with two relays, designated "low" relay and "high" relay, so called because one produces

low heat output and one high. At first call for heat, the first set of contacts closes in the thermostat and energizes the "low" relay which connects the heater unit to the neutral and a hot leg, supplying 120 volts to the heater. If the heat output is enough for the conditions of heat loss (depending upon outside temperature) the thermostat and low relay will cycle to maintain the desired temperature inside. If, however, the heater is not putting out enough heat to keep the room warm, the inside temperature will drop the extra degree to close the second set of contacts in the thermostat. This will energize the high relay, which disconnects the low relay and reconnects the heater unit for supply by two hot legs of the branch circuit—putting 240 volts across the heater and producing much higher heat output. Full heat output is provided at 240 volts. When operating at 120 volts, the unit produces only ¼ of its full rated output.

Location of any thermostat in the heated space is an important factor in effective control of the heaters. Since the heaters will operate only when the thermostat calls for heat, thermostats must be placed to accurately sense the comfort conditions in the room.

The mounting position of any thermostat must not be such that the thermostat receives direct radiation from lamps, appliances or other heat sources. And every thermostat should be protected from drafts which might circulate either in the stud space in which it is mounted or over its surface.

Demand or peak load controllers are used where sustained peak loads may be used by the utility to establish basic electrical energy rates or where demand charges are part of the rate structure. Such controls will either drop out part of the load when a given peak of connected load is reached or will convert operation to half voltage. Of course, the half-voltage method should not be used where motors may be damaged by lower-voltage operation.

Industrial Electric Heaters

FROM small electric soldering irons to large electric furnaces, industrial electric heating equipment is available in a great range of types, shapes and sizes. From the standpoint of operating principles, however, it can be divided into three categories: resistance heaters, high-frequency heaters, heat lamps.

Resistance Heaters

All resistance heaters operate on the same basic principle: when electric current flows through a conductor, the resistance of the conductor causes friction and heat is generated as a result. The heating elements consist of high-resistance (nickel chromium) wire embedded in insulating material, encased in tubing or other protective cover.

• **Immersion Heaters**—Many types of resistance heating devices are available for heating liquids in tanks, drums, kettles and various other metal or wooden containers.

There are two types of immersion heaters in common use for heating water, oil and alkaline solutions. One type has a threaded header and is installed through the side of tanks or containers. The other type is hung over the side of a container and down into the liquid to be heated.

This type is particularly applicable to wood or other non-metallic tanks.

Special lead-sheathed immersion heaters are available for nickel and copper plating and similar solutions. Cast lead types are used for chrome plating and sulphuric acid baths. Heaters used for melting soft metals (type metal, solder, babbitt and tin) consist of a steel sheathed tubular element cast into iron.

Immersion heaters for water have copper or alloy sheaths. Immersion heaters for oil have steel sheaths.

Typical applications of water-type immersion heaters are: water baths, cleaning tanks, stills, electric steam radiators, sterilizers and water heaters.

Non-circulating type oil immersion heaters are used for: oil heating, oil sterilizing, fuel oil preheating and alkaline baths. Circulating oil heaters are used in pipes or other vessels through which the oil is flowing.

• **Strip Heaters**—These heaters consist of a flat steel sheath in which the resistance wire element is encased in insulating material and stretched the length of the strip several times. Mounting slots are located at the ends of each strip to accommodate bolting of the heater.

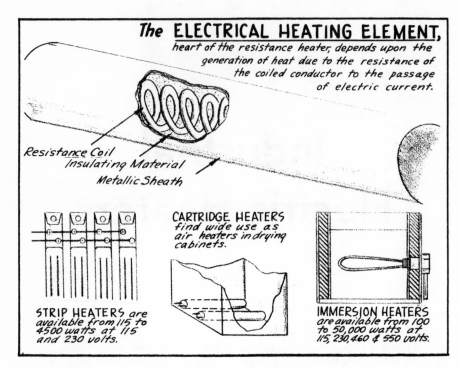

The ELECTRICAL HEATING ELEMENT,

heart of the resistance heater, depends upon the generation of heat due to the resistance of the coiled conductor to the passage of electric current.

Resistance Coil
Insulating Material
Metallic Sheath

STRIP HEATERS *are available from 115 to 4500 watts at 115 and 230 volts.*

CARTRIDGE HEATERS *find wide use as air heaters in drying cabinets.*

IMMERSION HEATERS *are available from 100 to 50,000 watts at 115, 230, 460 & 550 volts.*

Strip heaters are commonly applied in: drying ovens, compound tanks, incubators, melting pots, process machinery, platens, water baths, glue tables and pipe lines.

• **Finned Heaters**—These are essentially strip or tubular heaters with closely spaced radiating fins along the length of the element. The heat developed by the encased resistance wire is transferred to the fins.

Finned heaters find application in: blower-type electric unit heaters, air ducts with forced air circulation, car heaters, industrial processes requiring air blast heat for drying or baking, ovens and dryers.

• **Tubular Heaters**—These units consist of resistance heating element embedded in insulating material and enclosed by a metallic tubular sheath. Steel sheaths are used for heating air, hot plates, ironing machines and for heating liquids not harmful to steel.

In addition to immersion heating, tubular heaters can be used for contact heating of metal surfaces, casting into iron and aluminum, air heating and pipe heating. These heaters are also commonly used for heating ovens and contact heating of tanks.

• **Cartridge Heaters**—This type of heater consists of a resistance wire element wound on an insulator core. The element is connected to two terminals embedded in one end of the core, and the assembly is encased in a metal jacket.

Cartridge heaters are used for localized heating, such as: heating glue pots, compound pots, branding irons, platens and other devices where spots of heat are desired. The cartridge heater is snugly inserted in a hole of the same diameter as its sheath.

• **Radiant Heaters**—These units consist of a tubular resistance element mounted in an aluminum reflector. Used singly or in banks of multiple units, this type of heater produces infrared heat radiation for a wide range of drying, baking, curing, dehydrating and sealing applications.

• **Heating Cable**—This type of heating equipment today finds widespread industrial application. Basically, any electric heating cable consists of resistance wire

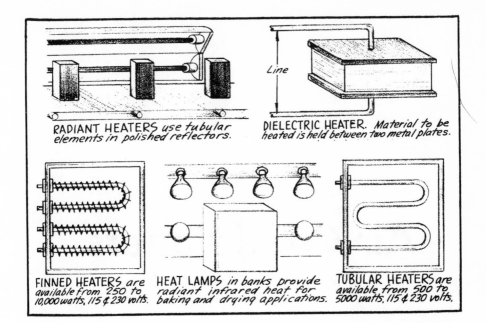

RADIANT HEATERS *use tubular elements in polished reflectors.*

DIELECTRIC HEATER. *Material to be heated is held between two metal plates.*

FINNED HEATERS *are available from 250 to 10,000 watts, 115 & 230 volts.*

HEAT LAMPS *in banks provide radiant infrared heat for baking and drying applications.*

TUBULAR HEATERS *are available from 500 to 5000 watts, 115 & 230 volts.*

in rugged protective insulating sheath of flexible material. Heating cable is made in a number of standard lengths which can be used singly or in series. Such cable is usually used in such applications as: along roofs and gutters to prevent accumulation of ice or snow; embedded in concrete of driveways, steps or sidewalks to eliminate ice or snow; wrapped around pipes to prevent freezing.

High-Frequency Heaters

Two types of high-frequency heating are in common use today: induction heating and dielectric heating.

In induction heating, any metallic object to be heated is placed in the high frequency magnetic field of an induction coil. The field induces rapidly alternating electric currents in the work, and heat is developed in the work as a result. This heating effect is quick acting.

A high frequency source of power is required to supply the induction coil. For large, easy-to-heat objects, rotating alternators with output frequencies from 1000 to 10,000 cycles per second can be used. Heating of small objects requires higher frequencies. Most electronic induction heating sources provide frequencies of 450,000 cycles per second.

Induction heating is used for: surface hardening of steel, brazing and soldering, annealing of brass and bronze and sintering powdered metals.

In dielectric heating, objects made of poorly conducting materials are placed in high frequency electrostatic fields between two plates. Molecules of the materials are pulled and pushed by the voltage of the electrostatic field, causing friction and developing heat as a result. Again a high frequency generator is used to supply power to the plates.

Typical applications of dielectric heating are: gluing, drying and curing of wood; drying and heat treating of textiles; processing of rubber; and treatment of foods.

Heat Lamps

The use of incandescent reflector type lamps as a source of electromagnetic radiation in the infrared region are used either singly or in banks for radiant heat processing in scores of baking, drying, curing and heating applications. Lamps can be arranged in any number of ways to make small or large ovens and heating tunnels.

Air Conditioners

AIR CONDITIONING is the process of controlling simultaneously the temperature, relative humidity, movement and quality of air in interiors used for human occupancy. Understanding of the subject requires familiarity with certain terms:

• **Condensation**—Process of changing vapor into liquid by extraction of heat.

• **Dehumidification**—Process of decreasing the amount of moisture in the air in a given space.

• **Evaporation**—Process of changing liquid into vapor by addition of heat.

• **Humidification** — Process of increasing the amount of moisture in the air in a given space.

• **Ventilation**—Process of supplying air to or removing it from any space.

Operation

Basically, all air conditioners are refrigerators, similar in components and principle of operation to the home food refrigerator. By the process of refrigeration, air conditioners take heat from an area to be air conditioned. The following sequence of operations typifies the refrigeration cycle in the air conditioning process:

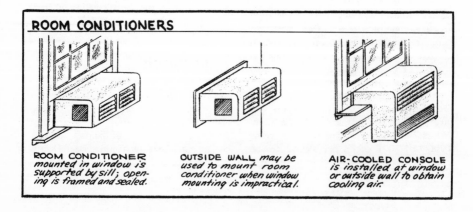

ROOM CONDITIONERS

ROOM CONDITIONER mounted in window is supported by sill; opening is framed and sealed.

OUTSIDE WALL may be used to mount room conditioner when window mounting is impractical.

AIR-COOLED CONSOLE is installed at window or outside wall to obtain cooling air.

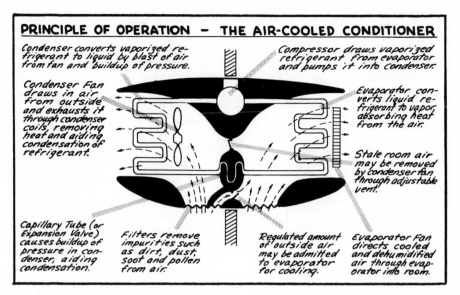

PRINCIPLE OF OPERATION — THE AIR-COOLED CONDITIONER

Condenser converts vaporized refrigerant to liquid by blast of air from fan and buildup of pressure.

Compressor draws vaporized refrigerant from evaporator and pumps it into condenser.

Condenser Fan draws in air from outside and exhausts it through condenser coils, removing heat and aiding condensation of refrigerant.

Evaporator converts liquid refrigerant to vapor, absorbing heat from the air.

Stale room air may be removed by condenser fan through adjustable vent.

Capillary Tube (or Expansion Valve) causes buildup of pressure in condenser, aiding condensation.

Filters remove impurities such as dirt, dust, soot and pollen from air.

Regulated amount of outside air may be admitted to evaporator for cooling.

Evaporator Fan directs cooled and dehumidified air through evaporator into room.

1. Air from the interior to be conditioned is drawn into the air conditioner by fan suction.

2. This air is passed over a grill of tubing called the "evaporator."

3. Within the evaporator, a refrigerant—a liquid with a very low boiling point—draws the heat out of the air, causing the refrigerant to change from liquid to vapor.

4. A motor driven compressor draws the vaporized refrigerant from the evaporator to a section called the "condenser."

5. Air or water passing over the condenser then picks up the heat from the vaporized refrigerant and carries this heat away, causing the refrigerant to condense back to liquid again.

6. The liquid refrigerant is returned to the evaporator where it again picks up heat from the air, is vaporized as a result and carries the heat to the condenser from which point it is carried off.

7. The cycle is continuous; the removal of heat from the air is a constant process.

Dehumidification may be accomplished in a number of ways, all based on condensation of moisture on the evaporator coils. Circulation is provided by fans. Filtering (by means of fiberglas, hair or paper filters) removes dust, dirt, soot and pollen from the air.

Basic electrical components of an air conditioner are as follows:

• **Compressor Motor** — The motor sealed within the compressor housing represents the major part of the electrical load. Compressor motors are rated at $\frac{1}{3}$-hp. and up.

• **Fan Motors** — Small motors are used to drive the condenser and evaporator fans. These represent almost negligible electrical load compared to the compressor motor.

• **Thermostats** — Some utility companies prohibit the use of thermostatic control of air conditioners in their areas, since frequent start and stop of conditioners would disrupt the electrical system, often causing flicker on lighting circuits.

Application

The following two types of unit air conditioners are unitized assemblies which accomplish the complete process of air conditioning:

• **Room Conditioner**—In capacities of less than $1\frac{1}{2}$ tons, this is the familiar window unit or the small air-cooled console unit which mounts against a window, neither of which requires ducts or water piping.

• **Self-contained Conditioner** — In

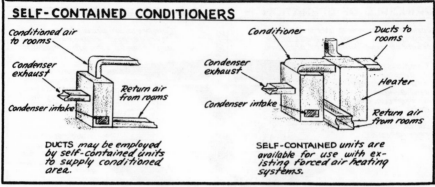

SELF-CONTAINED CONDITIONERS

Conditioned air to rooms

Condenser exhaust

Condenser intake

Return air from rooms

DUCTS *may be employed by self-contained units to supply conditioned area.*

Conditioner

Ducts to rooms

Condenser exhaust

Condenser intake

Heater

Return air from rooms

SELF-CONTAINED *units are available for use with existing forced air heating systems.*

capacities of 1½, 2, 3, 5, 7½ and 10 tons (some larger), this unit is usually applied in stores, restaurants, office areas and similar interiors which are too large to be handled economically by a window unit (or units). This unit can be used with or without distribution air ducts, can be adapted to use in conjunction with warm air furnaces. Most units of this type, particularly those above 3-ton rating, require water piping for water-cooled condensers. Water-cooled units are much more efficient than air-cooled units and offer consequent economies.

Since the primary function of an air conditioner is the transfer of heat, it is logical to rate units according to their capacity for removing heat from the space to be conditioned. Heat removal at the rate of 12,000 Btu. (units of heat) per hour is described as 1-ton of air conditioning capacity. Air conditioners are rated in tons of air conditioning capacity. By a coincidence of thermodynamic and electrical units, it so happens that the numerical value of a conditioner's rating in tons is the same as the numerical value of its compressor hp. rating. As a result, a 1-ton conditioner has a 1-hp. compressor, a 5-ton conditioner has a 5-hp. compressor, a ½-ton unit has a ½-hp. compressor, etc.

Selection of the proper type, size and number of units for any air conditioning job begins with calculation of the required cooling capacity of the equipment. For both room conditioners and self-contained units, the various manufacturers make available practical, easy-to-use cooling load estimate forms.

If the estimated cooling load is less than 1½ tons, a single room conditioner will satisfy the requirements. If it is more than 1½ tons, either multiple room units or one or more self-contained units must be selected. Generally, the selection of a single unit or a minimum number of units will be most economical. Typical considerations which might resolve the problem of number of units are as follows:

• Where the season of peak cooling is short, multiple units offer an advantage over the larger single unit: all of the units can be used during peak season, only as many as necessary need be used at other times.

• Where the air conditioning equipment must be installed outside of the conditioned area and connection to outside air is not possible, air-cooled units are ruled out. One or more self-contained conditioners must be used.

• Where available utilization voltage is 110-125, single-phase, only room type conditioners can be used. If the interior is rewired for 230 volts, single-phase (or 208 volts, 3-phase; 220 volts, 3-phase; 440-volts, 3-phase), self-contained conditioners can be used.

Installation

Without adequate wiring capacity, air conditioning cannot be safely and effectively applied. In any interior, the size of branch circuit wiring and feeder capacity must be judged on their ability to handle any proposed air conditioning load.

Direct-Current Generators

A DIRECT-CURRENT generator is a rotating machine for converting mechanical energy into direct-current electric power. A dc generator is driven by mechanical means, such as a steam engine, diesel engine or electric motor. Direct-current generators are rated in kilowatts; sizes range from a fraction of a watt to several thousand watts.

Operation

The basic electrical principle underlying operation of dc generators is called "generator action." If a length of wire is moved through a magnetic field in such a way that the wire cuts between the North and South poles producing the magnetic field, an electric voltage will be set up in the wire. Of course, the voltage set up in a single length of wire would be small. But if a large number of wires were connected end to end and rotated in the magnetic field, the voltage set up in the series of wire lengths would be equal to the sum of voltages set up in the individual lengths of wire. "Generator action" is the effect whereby the volt-age is set up in a moving conductor.

In dc generators, the conductors are wound into coils and mounted in slots on the armature or rotating element within the unit's housing. The ends of each coil of conductors are brought out to one end of the armature where they are connected to individual insulated bars in an assembly of such bars, which is mounted around the armature shaft. This assembly is called the "commutator," which is required to provide direct-current output from the generator. The wires that come out of the generator, one for the positive side of the dc output and one for the negative side, are connected within the generator to carbon brushes which ride on the commutator. The brushes pick up the electric current from the commutator and pass it to the circuit being fed by the generator.

The magnetic field required for generator action is produced in three ways:

• By permanent magnets, in which case the machine is called a magneto. Magnetos are generally small in size and capacity due to the limited strength of permanent magnets. Magnetos are used for various speed indicating devices.

- By electromagnets excited by a direct-current source independent of the generator itself. In such cases, magnetic pole pieces are parts of the stator (stationary housing of the generator in which the armature rotates). Each pole piece has one or two windings around it, which when energized by direct current will make the pole piece either a North or South magnetic pole. The windings on the pole pieces are called "field-coils" and when they are energized from a separate battery or a small generator, the term "separately excited" is used to describe the dc generator.

- By electromagnets excited by current from the generator's own armature. In these cases, the general arrangement of pole pieces and windings is the same as described in the previous case; but the field coils are connected to the brushes which pick up the output voltage of the armature, instead of being connected to an external source of direct current. Generators of this type are called "self-excited generators."

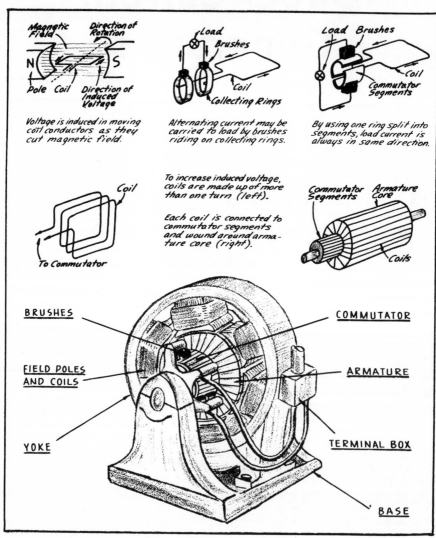

Voltage is induced in moving coil conductors as they cut magnetic field.

Alternating current may be carried to load by brushes riding on collecting rings.

By using one ring split into segments, load current is always in same direction.

To increase induced voltage, coils are made up of more than one turn (left).

Each coil is connected to commutator segments and wound around armature core (right).

BRUSHES

COMMUTATOR

FIELD POLES AND COILS

ARMATURE

YOKE

TERMINAL BOX

BASE

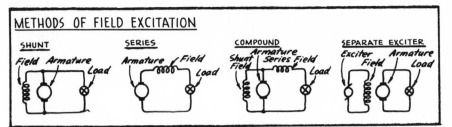

The self-excited generator is made in three types according to the arrangement of the field and armature connections:

- **Series generator**—In this type, one of the conductors which carries current from the commutator is brought directly out of the generator. The other conductor, which is also connected to a brush on the commutator, is then connected to the field coils. The field coils are in series and the end of the field coil windings becomes the other output conductor. The armature windings and the field coils are therefore in series.

- **Shunt generator**—The field coils of the shunt generator are connected across the armature terminals, in parallel instead of in series.

- **Compound generator**—This type of generator has two sets of field coils on the pole pieces. One set of field coils is arranged as in a series generator; the other set, as in a shunt generator.

Construction

Direct-current generators are made up of a more-or-less standard group of parts:

- **The yoke**—It is the cylindrical housing of the generator. It acts as a mechanical support for the pole pieces inside and for the end brackets which hold the bearings for the armature shaft. The yoke and the supporting feet of the generator make up "the frame" of the generator.

- **Pole pieces**—They are, with few exceptions, separate pieces attached to the inside of the yoke. The face of a pole piece is curved to conform to the curvature of the armature. The field coil is wrapped tightly around the pole piece, with the face of the pole extending toward the armature. The space between pole face and armature is the "air gap."

- **Field coils**—They conform to the shape of the pole pieces (circular or rectangular). The coils on the different pole pieces are connected in series, and the way the coil is wound determines the polarity of the pole Shunt field coils consist of many turns of fine wire and have a relatively high resistance. Series field coils consist of only a few turns of large size wire and have a low resistance.

- **Armature**—It is the rotating part of the generator. It is made up of four principal parts: the shaft, the iron core, the armature winding (conductors) and the commutator. The armature core is made up of punched laminations assembled in such a way as to provide slots either parallel to the shaft or at a slight angle with the shaft. The armature conductors are wound in coils and placed in the slots on the armature. Coil ends are soldered into the commutator segments. The armature conductors are held in place by bands of wire around the armature or by wedges driven into the slots.

- **Commutator**—It is a very important part of a dc generator. It consists of a large number of copper segments insulated from each other and from the armature shaft which they surround.

- **Brushes**—They are the rectangular blocks of carbon which are held in special holders to ride on the commutator, conducting direct current to or from the machine. The flexible copper cable on a brush is called a "pig-tail."

Application

Specs on typical generators for supplying dc power are as follows:

- **Speeds:** Generators are available for high speed (500 to 1800 rpm) and low speed (120 to 450 rpm) drivers.

- **Voltages:** 125, 250 and 600 volts.

Alternating-Current Generators

AN alternating-current generator is a rotating machine for converting mechanical energy into alternating-current electric power. As in the case of a dc generator, the rotating part of the machine is driven by a source of rotating power external to the machine itself. Alternating-current generators, also called "alternators," can be divided into three types according to the type of drive or prime mover. There are direct-connected engine-driven (steam or internal combustion) ac generators, water-wheel ac generators, turbine-driven ac generators.

Operation

Basically, the alternator operates on the principle of generator action—voltage is set up in coils of wire when they are moved in a magnetic field in such a way as to cut the magnetic lines of force. This is the same principle on which the dc generator operates. But there is a basic difference between dc generators and alternators in accomplishing the relative motion between coils and field.

• In a dc generator, the magnetic field is set up by placing electro-magnetic field poles around the inside of the housing or stator. Coils are placed on the rotor so that when the rotor rotates, the coils are cutting across the magnetic lines of force set up by the poles and voltage is produced in the rotor coils. By the use of a commutator and brushes, voltage is taken from the rotor.

• In an ac generator, the electro-magnetic field—the source of the magnetic lines of force—is generally placed on the rotor and the coils in which the voltage is to be generated are mounted around the inside of the stator. In this case, the rotating field sweeps the lines of magnetic force across the stationary coils and produces the voltage in them.

Alternators with stationary coils (or stationary armature, as it is sometimes described) have distinct advantages over alternators with rotating coils and are the most common type. If the generator coils are placed on the rotor, two or more slip rings must be used for conducting the ac power to the external circuit which the generator feeds. Such rings are a frequent source of trouble, particularly at the high voltage put out by standard alternators. If stationary coils are used, the generated ac power can be taken directly from the coil leads to output bus bars. Also, if the coils are on the rotor, it is more difficult to properly insulate them, and they are subjected to heavy centrifugal forces and vibrations.

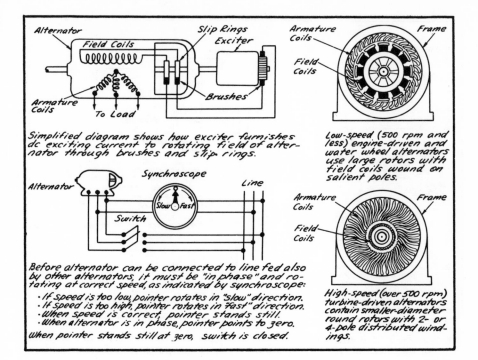

Simplified diagram shows how exciter furnishes dc exciting current to rotating field of alternator through brushes and slip rings.

Before alternator can be connected to line fed also by other alternators, it must be "in phase" and rotating at correct speed, as indicated by synchroscope:
· If speed is too low, pointer rotates in "slow" direction.
· If speed is too high, pointer rotates in "fast" direction.
· When speed is correct, pointer stands still.
· When alternator is in phase, pointer points to zero.
When pointer stands still at zero, switch is closed.

Low-speed (500 rpm and less) engine-driven and water wheel alternators use large rotors with field coils wound on salient poles.

High-speed (over 500 rpm) turbine-driven alternators contain smaller-diameter round rotors with 2- or 4-pole distributed windings.

When the electro-magnetic field is placed on the rotor, as in most commercial alternators, the direct current necessary to energize the field must be conducted to and from the rotor by the use of slip rings. However, the voltage seldom exceeds 250 volts and the amount of power involved is so small as to cause no particular difficulties.

• **Windings**—In general, windings for alternators are the same as for dc generators. Single-phase windings are almost never used in alternators. Generators for single-phase railroad supply are the exceptions. Usually, alternator windings are arranged to provide three-phase output of ac electric power. Such an arrangement is most efficient and economical. Single-phase power can be taken from a three-phase alternator. Actually, three-phase windings in alternators are simply three single-phase windings properly spaced in the stator slots.

• **Field Excitation**—Generally, current is supplied to the field windings on the rotor by an individual exciter which may be driven directly or through a gear reduction. Excitation voltage varies between 120 and 250 volts. Excitation voltage is also often obtained from exciters coupled directly to the alternator shaft or from a bank of batteries.

Construction

Engine-driven, water-wheel and turbine-driven alternators differ in design.

• Engine-driven (gasoline or diesel engine) alternators operate at very low speeds. On such units, a separate or integral flywheel is used to provide constant speed of rotation, overcoming the rotational impulses due to piston action.

• Water-wheel alternators vary in speed over a wide range—from 60 to 500 rpm, the lower speeds being used at the lower heads of water. These units are designed to operate at twice their rated speed: they are made in both vertical and horizontal types, with the vertical types more commonly used.

• Turbine-driven alternators generally run at high speeds—750 to 3,600 rpm. This type of alternator is invariably of the horizontal construction.

In all of these alternator types, there are certain basic parts common to all:

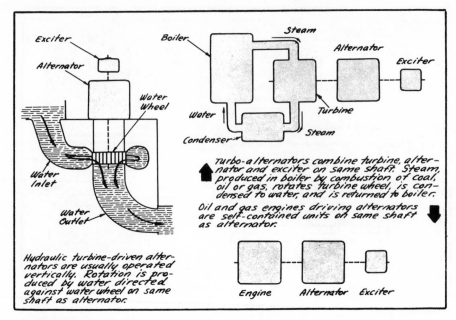

Turbo-alternators combine turbine, alternator and exciter on same shaft. Steam, produced in boiler by combustion of coal, oil or gas, rotates turbine wheel, is condensed to water, and is returned to boiler.

Oil and gas engines driving alternators are self-contained units on same shaft as alternator.

Hydraulic turbine-driven alternators are usually operated vertically. Rotation is produced by water directed against water wheel on same shaft as alternator.

Engine   Alternator Exciter

• **Stator**—This is the stationary member of the alternator and almost always is the armature—the coils in which the voltage is generated. The stator is constructed of steel laminations dovetailed or bolted to the frame and arranged with parallel slots to accommodate the coils. The frame or housing may be a hollow box casting, or it may consist of steel plates between which the stator laminations are held. Perforations in the steel laminations or other longitudinal ducts are provided in the stator for ventilation.

• **Coils**—These are wound in the slots in the laminated stator. Insulation on the coils is divided into three common classes. Class A insulation consists of paper or cambric impregnated with varnishes or fillers; it has a limiting operating temperature of 100° C. Class B insulation consists of mica or fiber glass; it can operate at a limiting temperature of 120° C. Class H insulation is the newer silicone insulation; it has an operating temperature limit of 160° C.

• **Ventilation**—Slow speed alternators are easily ventilated by providing ducts for air circulation. Turbine-driven (high speed) alternators, however, are not so easily ventilated. Nearly all turbine-driven alternators are ventilated by hy-drogen gas in enclosures with leak proof seals where the shaft comes out.

• **Rotor**—This is the rotating part of the alternator and consists of the electromagnetic poles which produce the magnetic lines of force. For slow speed alternators, rotors are made of laminated steel punchings riveted together and are called "salient-pole" rotors. Rotors for turbine-driven alternators are cylindrical solid steel forgings.

Applications

Alternating current generators are available for producing a wide range of power outputs in many applications.

• **Rating**—An electric machine is rated according to its temperature rise. The temperature rise is caused by the power losses in the machine which show up as heat. The current drawn from an alternator determines the heat rise. For this reason, alternators are rated in kva—kilovolt-amperes — 1,000 times their rated voltage times the maximum current they can carry before getting dangerously hot. The power factor is usually given with the kva rating. Power factor times kva gives the kilowatt rating of the alternator.

Direct-Current Motors

A DIRECT-CURRENT motor is a rotating machine for converting dc electrical energy into mechanical energy—the reverse of the action performed by a dc generator. In construction, appearance and operation, they are similar. However, where the generator produces electric power due to rotation, the motor produces rotating power due to electrical action and reaction within the machine.

Operation

The principle underlying operation of every dc motor is called "motor action." Like generator action, it involves current, magnetism and motion. When a wire carrying current is placed in a magnetic field, a force is exerted on the wire, moving it through the magnetic field.

Operation of a typical dc motor is an extension of the foregoing simple action. Instead of just a length of wire, the motor contains many coils of wire wound on a cylindrical rotor or armature on the shaft of the motor. The wire coils are so wound that the coil sides are embedded in the armature parallel to the motor shaft.

The second requirement is creation of a magnetic field. This is done by placing electromagnetic poles within the motor housing so that when the armature is placed in the housing, the wires embedded in it will be in the field of magnetic lines of force, which pass from a north pole to a south pole.

The third condition is that the armature be free to rotate within the housing—that the shaft be supported at both ends by bearing brackets and that the armature be free of contact with the pole pieces or other parts. If, now, electric current is passed through the coils such that current always flows in one direction when wires are under the north pole and in the opposite direction when wires are under the south pole, the forces acting on the wires will cause rotation of the armature.

The trickiest part of dc motor operation is the action whereby conductors on the armature have current flowing through them in one direction when they are under a north pole and current flowing through them in the opposite direction when they are under a south

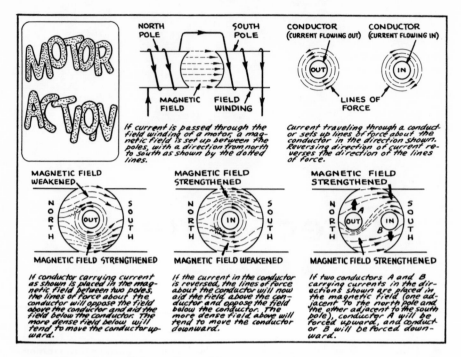

MOTOR ACTION

NORTH POLE — SOUTH POLE — MAGNETIC FIELD — FIELD WINDING

If current is passed through the field winding of a motor, a magnetic field is set up between the poles, with a direction from north to south as shown by the dotted lines.

CONDUCTOR (CURRENT FLOWING OUT) — CONDUCTOR (CURRENT FLOWING IN) — LINES OF FORCE

Current traveling through a conductor sets up lines of force about the conductor in the direction shown. Reversing direction of current reverses the direction of the lines of force.

MAGNETIC FIELD WEAKENED — NORTH — OUT — SOUTH — MAGNETIC FIELD STRENGTHENED

If conductor carrying current as shown is placed in the magnetic field between two poles, the lines of force about the conductor will oppose the field above the conductor and aid the field below the conductor. The more dense field below will tend to move the conductor upward.

MAGNETIC FIELD STRENGTHENED — NORTH — IN — SOUTH — MAGNETIC FIELD WEAKENED

If the current in the conductor is reversed, the lines of force about the conductor will now aid the field above the conductor and oppose the field below the conductor. The more dense field above will tend to move the conductor downward.

MAGNETIC FIELD STRENGTHENED — NORTH — OUT — IN — SOUTH — MAGNETIC FIELD STRENGTHENED

If two conductors A and B carrying currents in the directions shown are placed in the magnetic field (one adjacent to the north pole and the other adjacent to the south pole), conductor A will be forced upward, and conductor B will be forced downward.

---

pole. The action which produces this change in direction of current flow is called "commutation."

Commutation is necessary in the case of a dc motor because the electric power supplied to the motor has the characteristic of continuous flow of current in one direction (direct current). The conductors supplying the motor are connected to rectangular blocks of carbon or graphite, called "brushes." These brushes are held in brush-holders which are secured to the stationary frame of the motor. The brushes make contact with the commutator, an assembly of copper bars mounted compactly around the motor shaft.

Each of the individual bars in the commutator assembly is connected to ends of coils of wire set in the slots in the armature. The brushes, positive and negative, are so spaced around the commutator that when wires on the armature come under a north pole, the bars to which they are connected are between a pair of brushes with a definite direction of current feed. Now when the same wires have rotated under a south pole, the bars connected to the

wires are between brushes with opposite direction of current feed.

To provide the magnetic field necessary for motor action, electromagnetic north and south poles are used. The windings on these poles are called "field coils." The difference between a north and south pole is effected by feeding current in opposite directions to the field coils. The way field coils are wound and connected to the supply circuit determines the characteristics of the motor.

There are three basic classes of direct-current motors:

• **Shunt Motor**—In a shunt motor, the field coils on the pole pieces are connected in series with each other and the entire group is connected across the armature terminals, in shunt (or parallel) with the armature.

• **Series Motor**—In this class of motor, the field coils are connected in series with each other and the group is connected in shunt with the armature.

• **Compound Motor**—This class of motor includes all dc motors in which two windings are used on each pole piece—one series winding and one shunt winding. The series windings are all connected in series and the group of them connected in series with the armature; the shunt windings are connected in series with each other and the group connected in shunt with the armature.

Application

Direct-current motors are rated in terms of voltage, speed and power output. Standard voltage ratings of dc motors are as follows: 6 volts for automobiles; 12 volts for trucks and buses; 110-125 volts for small general purpose motors; 220-230 volts for power applications; 500 and 1500 volts for electric railway service.

Rated speed of dc motors varies widely. Because shunt and compound wound dc motors are so well adapted to speed adjustment, many motors are designed as adjustable-speed units. Such motors for 10 to 150 hp have a speed range of three or four to one, with speeds as low as 300 rpm and up to 1600 rpm. Constant-speed dc motors have speeds which correspond to standard speeds of ac motors: 575, 850, 1150 and 1750 rpm. Power ratings of dc motors are given in horsepower. The current-rating of a motor provides the basis for selection of feeders and protective devices.

Speed characteristics provide the basis for classification of dc motors:

• A constant-speed motor will maintain its speed constant for any load.

• An adjustable-speed motor allows speed variation through wide limits, but the speed for any adjustment is constant under varying load.

• A variable-speed motor reduces its speed as required turning force increases.

The shunt motor is essentially a constant-speed and adjustable speed motor; it is best used for constant-speed service—lathes, grinders, etc. The series motor is a variable-speed unit used extensively for cranes, hoists, elevators and locomotives. In the fractional hp sizes, the series dc motors are usually of the "universal" type, allowing use on ac as well as dc. Such small units are used in fans, vacuum cleaners, sewing machines, hair driers, etc.

The compound-motor may be of the constant, adjustable or variable type depending upon the arrangement of field coils. Typical applications of compound motors are: rolling mill drives, power shears, power fans and other applications requiring particular motor speed characteristics.

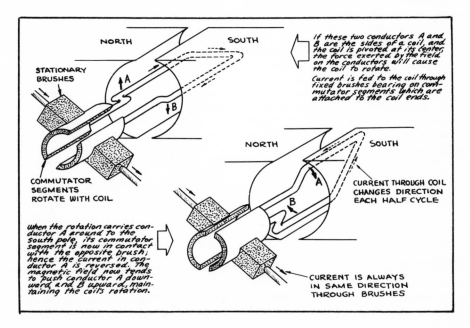

STATIONARY BRUSHES

NORTH   SOUTH

↑A

↓B

COMMUTATOR SEGMENTS ROTATE WITH COIL

If these two conductors A and B are the sides of a coil, and the coil is pivoted at its center; the force exerted by the field on the conductors will cause the coil to rotate.

Current is fed to the coil through fixed brushes bearing on commutator segments which are attached to the coil ends.

NORTH   SOUTH

A

B

CURRENT THROUGH COIL CHANGES DIRECTION EACH HALF CYCLE

When the rotation carries conductor A around to the south pole, its commutator segment is now in contact with the opposite brush; hence the current in conductor A is reversed. The magnetic field now tends to push conductor A downward and B upward, maintaining the coil's rotation.

CURRENT IS ALWAYS IN SAME DIRECTION THROUGH BRUSHES

Alternating-Current

Motors

ALTERNATING-CURRENT electric motors are rotating machines for converting alternating-current electric power into mechanical power. From fractional horsepower sizes to units rated at thousands of horsepower, alternating-current, or simply ac, motors are available in a wide variety of specific types for particular applications. Although dc motors are better suited to applications requiring control of motor speed, ac motors find much wider and general use than dc motors throughout industry today.

Operation

As in the case of dc motors, the basic principle called "motor action" underlies operation of all ac motors. According to this principle, if electric current flows through a wire while it is in and perpendicular to the lines of magnetic field, a force will be exerted, moving it across the field. Of course, an actual motor is an extension and elaboration on this fundamental idea. Instead of one, many current-carrying wires are made to react with a magnetic field, thereby developing usable rotating power.

In ac motors, the establishment of a rotating part in a stationary housing in such a way as to produce interaction between a magnetic field and current-carrying conductors is done in many ways, depending upon the specific types of motors.

Single-Phase Motors

Single-phase ac motors are supplied by only two wires, between which a voltage exists and alternates sinusoidally —for purposes of the discussion here, a 60-cycle single-phase voltage. Single-phase motors are available from fractional horsepower units to units rated over 25 hp. Typical voltages for single-phase motors are: 110, 220, 440, 550.

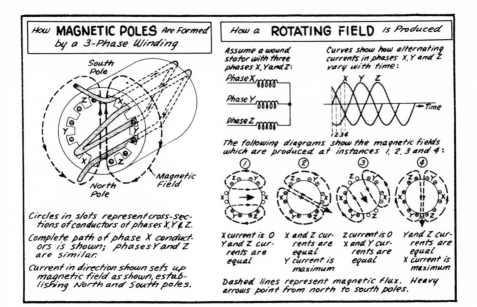

How MAGNETIC POLES Are Formed by a 3-Phase Winding

Circles in slots represent cross-sections of conductors of phases X, Y & Z.

Complete path of phase X conductors is shown; phases Y and Z are similar.

Current in direction shown sets up magnetic field as shown, establishing North and South poles.

How a ROTATING FIELD Is Produced

Assume a wound stator with three phases X, Y and Z:

Phase X
Phase Y
Phase Z

Curves show how alternating currents in phases X, Y and Z vary with time:

The following diagrams show the magnetic fields which are produced at instances 1, 2, 3 and 4:

(1)	(2)	(3)	(4)
X current is 0 Y and Z currents are equal	X and Z currents are equal Y current is maximum	Z current is 0 X and Y currents are equal	Y and Z currents are equal X current is maximum

Dashed lines represent magnetic flux. Heavy arrows point from north to south poles.

Universal Motors

This type of single-phase motor for use on 110 and 220 volt circuits is a special adaptation of the series connected dc motor and can be used as well on dc as on ac. Made in fractional hp sizes, the motor is constructed like the dc series motor and contains field windings on the stator within the frame, an armature with the ends of its windings brought out to a commutator at one end and carbon brushes which are held by the motor's end plate in position which allows their proper contact with the commutator.

When current is applied to a universal motor, either ac or dc, the current flows through the field coils and the armature windings in series. The magnetic field set up by the field coils in the stator react with the current-carrying wires on the armature producing rotation.

Universal motors are used on household appliances such as sewing machines, vacuum cleaners, electric fans, etc.

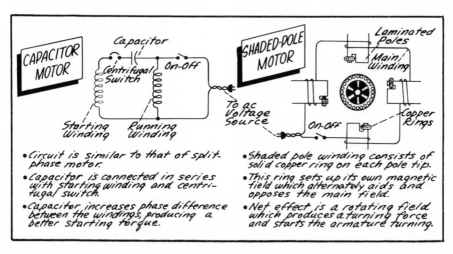

CAPACITOR MOTOR

- Circuit is similar to that of split-phase motor.
- Capacitor is connected in series with starting winding and centrifugal switch.
- Capacitor increases phase difference between the windings, producing a better starting torque.

SHADED-POLE MOTOR

- Shaded pole winding consists of solid copper ring on each pole tip.
- This ring sets up its own magnetic field which alternately aids and opposes the main field.
- Net effect is a rotating field which produces a turning force and starts the armature turning.

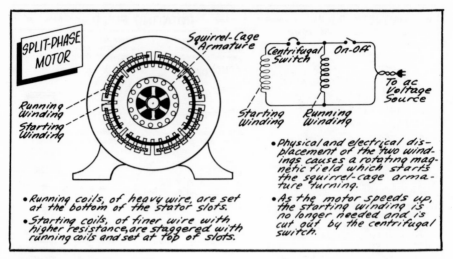

SPLIT-PHASE MOTOR

Squirrel-Cage Armature

Centrifugal Switch

On-Off

To ac Voltage Source

Starting Winding

Running Winding

Running Winding

Starting Winding

- Running coils, of heavy wire, are set at the bottom of the stator slots.
- Starting coils, of finer wire with higher resistance, are staggered with running coils and set at top of slots.

- Physical and electrical displacement of the two windings causes a rotating magnetic field which starts the squirrel-cage armature turning.
- As the motor speeds up, the starting winding is no longer needed and is cut out by the centrifugal switch.

Shaded-Pole Motors

These are small, 1/100 to 1/20 hp, ac motors used for applications requiring very low turning force (torque) to start up, such as small fans and blowers. These motors have no commutators on their armatures and no brushes. They get their name from the auxiliary or "shading" coil which is set in the field poles to provide initial turning force.

Split-Phase Motors

These are fractional hp units which use an auxiliary winding on the stator to get the motor going and up to proper speed of rotation. These motors are used in such devices as washing machines, oil burners, pumps, etc.

A split-phase motor consists of a housing, a laminated iron core stator with embedded windings forming the inside of the cylindrical housing, a rotor which is made up of copper bars set in slots in an iron core and connected to each other by copper rings around both ends of the core, end plates which are bolted to the housing and contain the bearings which support the rotor shaft, and a centrifugal switch inside the housing. The type of rotor described is called a "squirrel cage" rotor, from the resemblance which the configuration of copper bars bears to an actual cage. There are no wires wound on such a rotor. The centrifugal switch serves to open

the circuit to the starting winding when the motor comes up to speed.

Operation of a split-phase motor is as follows: current is applied to the stator windings, both the main winding and the starting winding which is in parallel with it through the centrifugal switch; the two windings set up a rotating magnetic field; the rotating field sets up a voltage in the bars of the squirrel cage rotor, and because these bars are shorted at the ends of the rotor, current flows through the rotor bars; the current-carrying rotor bars then react with the magnetic field to produce motor action When the rotor is turning fast enough, the centrifugal switch cuts out the starting winding which is no longer needed.

Capacitor Motors

Single-phase ac capacitor motors are made in sizes ranging from fractional hp to 15 hp. They are widely used in such devices as refrigerators, compressors, washing machines and oil burners. Construction of these motors is similar to that of the split-phase motors, with the addition of a capacitor (often called "condensers," particularly in radio usage) in series with the starting winding. The effect of the capacitor is to provide higher starting torque with lower starting current than the split-phase motor. The capacitor is mounted either in or on the motor housing. Capacitors are invariably tubular in shape, about 2 or 3 inches long, with connecting wire leads coming out of their body.

There are two types of capacitor motors as follows:

Capacitor-start motors utilize the capacitor only for starting; it is out of the circuit while the motor is running. These motors have the same parts as split-phase units. Operation is also similar. However, the capacitor in series with the starting winding produces stronger and quicker starting action. When the motor comes up to about 75 per cent of full speed, the centrifugal switch opens the starting circuit.

Capacitor start-and-run motors keep the capacitor and starting winding in parallel with the running winding at all times. These motors are quiet and smooth.

Repulsion-Type Motors

These single-phase motors are divided into 3 groups; repulsion-start, induction-run motors; repulsion motors; and repulsion-induction motors. Construction features common to all repulsion motors are:

- A stator which generally has one winding similar to the running winding on a split-phase motor.
- A rotor which consists of a slotted core with embedded windings connected to a commutator at one end of the rotor.

- End plates in each of which a bearing is mounted to support the rotor shaft.
- Carbon brushes are fitted in holders and ride on the commutator to provide a path for current flow between commutator segments and each brush.
- Brush holders supported in the motor, either on the front end plate or on the rotor shaft.

Details on each of the groups of repulsion-type motors are as follows:

Repulsion-start, induction-run motors range from 1/10 horsepower to 20 horsepower. They have high starting torque and constant speed characteristics. Such motors are found in large commercial refrigerators, pumps, compressors and other rotating machines with high starting torque requirements.

There are two designs of repulsion-start, induction-run motors. One, called the "brush-lifting" type, operates in such a way that current induced in the rotor winding by the energized stator can flow through the windings through the brushes which are in contact with the commutator and are connected to each other. When the motor comes up to 75 per cent of full speed, a centrifugal mechanism lifts the brushes from the commutator and short circuits the commutator segments (connects them all together). The motor then operates as a squirrel-cage induction motor until it is stopped. In the other type of repul-

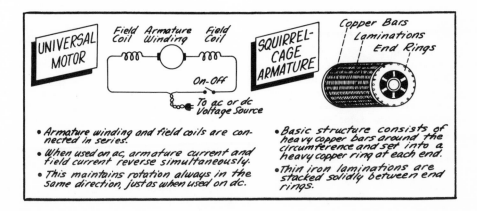

UNIVERSAL MOTOR

Field Coil | Armature Winding | Field Coil

On-Off

To ac or dc Voltage Source

- Armature winding and field coils are connected in series.
- When used on ac, armature current and field current reverse simultaneously.
- This maintains rotation always in the same direction, just as when used on dc.

SQUIRREL-CAGE ARMATURE

Copper Bars | Laminations | End Rings

- Basic structure consists of heavy copper bars around the circumference and set into a heavy copper ring at each end.
- Thin iron laminations are stacked solidly between end rings.

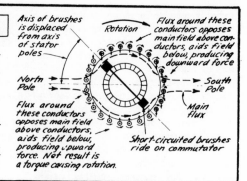

REPULSION MOTORS

REPULSION START - INDUCTION RUN
- Centrifugal device shorts commutator segments at about 75% full speed.
- Armature then acts as squirrel-cage rotor.
- Centrifugal device may also lift brushes during operation.

REPULSION
- Does not have centrifugal device.

REPULSION-INDUCTION
- Does not have centrifugal device.
- Has squirrel-cage winding in addition to distributed winding.

Axis of brushes is displaced from axis of stator poles

Rotation

Flux around these conductors opposes main field above conductors, aids field below, producing downward force

North Pole

South Pole

Flux around these conductors opposes main field above conductors, aids field below, producing upward force. Net result is a torque causing rotation.

Main flux

Short-circuited brushes ride on commutator

sion-start, induction-run motor, called the "brush-riding" type, the brushes remain in contact with the commutator at all times, even after the centrifugal shorting device has short circuited the commutator. Again, the motor runs as a squirrel-cage induction motor as soon as the motor speed is sufficient.

Repulsion motors differ from repulsion-start, induction-run motors in that they have no centrifugal mechanism and the brushes are always in contact with the commutator.

The repulsion motor has high starting torque, and it offers variable speed characteristics.

Repulsion-induction motors are almost identical to repulsion motors in external appearance. But the repulsion-induction motor has a squirrel-cage winding on the rotor in addition to the regular winding. It has no centrifugal mechanism and the brushes always ride on the commutator. Made in sizes up to about 10 hp., the repulsion-induction motor has high starting torque.

Three-Phase Motors

These motors are designed for application on 3-phase, 60-cycle, ac supply lines. They are available in a range of sizes from fractional horsepower to thousands of horsepower. Three-phase motors are made for just about every standard voltage and frequency. Typical examples of their use are: industrial machinery, machine tools, elevators, pumps, compressors, hoists, blowers and cranes.

Induction Motors

The most common types of 3-phase motors are the induction motors. There are two types of 3-phase induction motors: the squirrel-cage induction motor and the wound-rotor induction motor.

Construction of all 3-phase induction motors is basically the same. All consist of a steel housing or frame with a laminated and slotted iron core around its inside and windings of insulated conductors embedded in the core. This whole assembly is, of course, the stator. The rotor is the rotating part of the motor and is a shaft-mounted assembly which is supported at its shaft ends by bearings set in the end plates (or end bells or bearing brackets) of the motor. The end plates form the ends of the housing.

The difference between the two types of 3-phase induction motors is in the construction of the rotor. In the squirrel-cage induction motor, the rotor is made up of heavy copper bars fitted into slots, parallel to the shaft, in a laminated iron cylindrical core. At each end of the rotor, the bars are connected together by circular copper rings to which the bar ends connect. The whole core assembly is pressed onto the shaft. In the wound-rotor or slip ring induction motor, the rotor core has a winding of conductors set in the slots and connected to three slip rings (circular metal bands mounted on the shaft and insulated from each other) at one end of the rotor. Brushes which contact the slip rings are mounted in holders on the

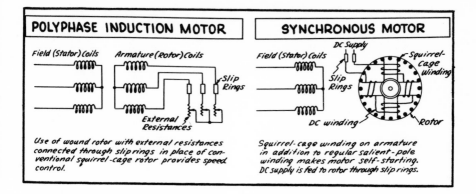

POLYPHASE INDUCTION MOTOR

Field (Stator) Coils   Armature (Rotor) Coils

Slip Rings

External Resistances

Use of wound rotor with external resistances connected through slip rings in place of conventional squirrel-cage rotor provides speed control.

SYNCHRONOUS MOTOR

DC Supply

Field (Stator) Coils

Slip Rings

Squirrel-Cage Winding

DC winding

Rotor

Squirrel-cage winding on armature in addition to regular salient-pole winding makes motor self-starting. DC supply is fed to rotor through slip rings.

inside of one of the end plates. The brushes are connected to external resistor grids which are used to limit the amount of current flow in the rotor winding, to control torque characteristics and to provide speed control.

The stators of the two types of motors are the same. The coils in the slots are connected to form three separate windings, one for each of the three phases of the supply line. Stator windings may be connected in either of two ways: delta, or star (also called "wye" connected). In either case, only three external connections are made to the supply circuit.

The squirrel-cage induction motor offers high reliability and efficiency at constant speed, in a wide range of torque and starting-current ratings. The wound-rotor induction motor offers limited speed control and speed adjustments under fluctuating load with low starting current and reduced efficiency.

Synchronous Motors

Three-phase motors of this type are made only in the larger sizes, from about 20 to several thousand horsepower. They are used for power factor correction to offset the poor power factor of a heavy concentration of induction motors. They are used for exact slow-speed drives and for maximum efficiency on continuous loads above 75 hp.

Synchronous motors are actually synchronous alternators operating as motors. A three-phase line feeds current to the stator and dc is supplied to the rotor. Reaction between magnetic fields causes the rotor to rotate at synchronous speed with the stator rotating magnetic field. As a result, the synchronous motor operates at constant speed. As in the case of the synchronous alternator (ac generator), the three-phase winding may be on the rotor, in which case the stator is energized with dc.

Wires and Cables

WIRES and cables are the most common types of conductors used to carry electric current throughout all types of circuits and systems. And wires and cables are available in a very wide variety of types and constructions suited to many different applications. Differences include size, insulation, outer jacketing, number of conductors and makeup of conductors. In any particular applications, wire or cable must essentially have sufficient mechanical strength to protect it against damage, insulation to keep the conductor(s) from improper contact with anything external to the circuit and adequate cross-sectional area to assure easy flow of the particular value of current.

Terminology

Definitions of common terms used in reference to wires and cables serve to clarify discussion:

• **A wire** is a single, solid length of drawn metal (generally copper, sometimes aluminum and for special purposes, other metallic compositions). A wire is specifically the fine metal rod itself, although when a wire is covered with insulation the combination is referred to as wire.

• **A conductor** may be a wire or a number of uninsulated wires twisted together. In either case, it represents a single path for conduction of electric current. Of course, bus bars are also conductors.

• **A stranded conductor** is a conductor made up of a number of uninsulated wires twisted or braided together.

• **A cable** may be either a single-conductor cable or a multi-conductor cable. The single-conductor cable is a single stranded conductor and is the

COPPER *is the most universally used metal for wire and cable due to its low resistance to the flow of current, the ease with which it can be bent and soldered, and its low cost.*

ALUMINUM, *due to its higher resistance, has less carrying capacity than copper but is extensively used because of its weight (only 30% as heavy as copper) and because it is plentiful. For bus bars, its large cross-section provides greater surface for radiation.*

SILVER, *although it is the best conductor of current, has only limited small-scale uses because of its high cost.*

simplest type of cable. In the case of the multi-conductor cable, the individual conductors in the cable may be solid or stranded conductors but they are always insulated from each other. The assembly of individual conductors in the multi-conductor cable may or may not have overall outer insulating cover. In some catalogs, a lead-covered, heavily-insulated solid wire is called a cable. The general term "cable" is frequently used to describe any large size conductor.

• **A stranded wire** is a number of small size wires, not insulated from each other, twisted together to form a single wire. This is distinguished from the stranded conductor called a cable only by the small size of the stranded wire.

• **A cord** is a small type of cable, made up of two or more individual conductors of stranded wire, with a common outer protective jacket.

In general, wires and cables are described according to size and insulation. Of course, full description of many cables involves specific construction characteristics. Size of wires and cables is expressed in terms of the American Wire Gage (AWG), known also as the Brown & Sharpe (B&S) gage. Insulations are basically divided into 600-volt insulations (for use in circuits of voltages up to 600) and high-voltage insulations.

In the United States, copper and aluminum wires and cables used in the electrical industry are designated according to the sizes of the American Wire Gage. Size designations run from #44, the smallest size, to #4/0, the largest size in the AWG. Sizes larger than #4/0 are expressed in circular mils, units of cross section area. As size increases from #44 to the larger sizes, the designation after #1 is #1/0 (or just 0). Then comes 2/0, 3/0 and 4/0 (00, 000 and 0000). The circular mil sizes start with 250,000 circular mils, commonly called 250 MCM (thousand circular mils). The largest circular mil size is 5,000,000 (or 5,000 MCM). In practical usage, 2,000 MCM represents the maximum size.

In electrical circuits for light, heat and power, #14 is the smallest size of wire allowed by the National Electrical Code. Smaller sizes of wire and cable are used in signal, communication and control circuits. Tables 1 and 2, Chapter 10 of the National Electrical Code, give the current-carrying capacities of different sizes of conductors with different types of insulations. Table 1 covers conductors in raceway, cable or direct burial; Table 2 covers conductors in free air. Thorough

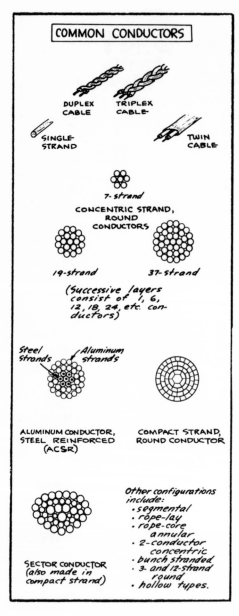

COMMON CONDUCTORS

DUPLEX CABLE

TRIPLEX CABLE

SINGLE-STRAND

TWIN CABLE

7-strand

CONCENTRIC STRAND, ROUND CONDUCTORS

19-strand

37-strand

(Successive layers consist of 1, 6, 12, 18, 24, etc. conductors)

Steel strands

Aluminum strands

ALUMINUM CONDUCTOR, STEEL REINFORCED (ACSR)

COMPACT STRAND, ROUND CONDUCTOR

SECTOR CONDUCTOR (also made in compact strand)

Other configurations include:
· segmental
· rope-lay
· rope-core
  annular
· 2-conductor
  concentric
· bunch stranded
· 3- and 12-strand
  round
· hollow types.

degree of covering used. The first class is the bare conductor, widely used in electrical circuits. Such conductors have no covering on them. The most common use of bare conductors is in electrical overhead transmission lines —the familiar high-tension line. The second class is covered, but not insulated, conductors. Such coverings provide protection against weather and resistance to heat. Typical covered cables are used for outdoor, low-voltage overhead circuits and aerial distribution and for exposed interior wiring. The third class covers insulated conductors, which have an insulating covering over the metallic conductor. This insulation serves to electrically isolate the conductor within and to allow grouping of such conductors close together without the voltage between the conductors breaking down the insulation and allowing current flow from one to the other. An additional covering may be used over the insulation to add mechanical strength and protection against weather, dampness, abrasion, etc.

By far, the majority of insulated wires and cables are made of copper. However, the use of insulated wires and cables of aluminum is continually growing. Particularly in the larger sizes, used for feeders in distribution systems, aluminum cables are competitive with copper in many cases, today. Of course, aluminum or steel conductors are widely used for aerial electrical construction. Steel conductors and steel and copper conductors are used for long-span aerial lines. Telephone and telegraph lines are typical of such applications. Aluminum cable, steel-reinforced (ACSR) is a well-known aerial cable.

Bare cables are necessarily single-conductor cables. Insulated cables may have one, two, three or even more individually-insulated conductors within the cable assembly. In general, single-conductor cables are more commonly used than multi-conductor cables, particularly in interior wiring.

familiarity with these tables is essential to full understanding and facility in effective application of wire and cables.

A basic classification of wires and cables can be made according to the

S ELECTION and application of electrical wires and cables is based primarily on size of conductors and type of insulation and covering. The sizes of conductors are given in the Tables in the back of the National Electrical Code. The larger the size, the higher the allowable current which the conductor may be used to carry. Selection of cable or wire according to current-carrying capacity is made on the basis of the characteristics of electrical circuits and the rating of the load device to be supplied. Design calculations and the provisions of the National Electrical Code determine the necessary size and type of wire or cable to be used in any particular case.

The Code Tables give the allowable current ratings of sizes of conductors with various types of insulations. The values given for current in conductors are based on temperature rise due to heating effect of current flow. The Tables cover conductors in conduit, in cable or in open air. Typical current ratings of conductors are as follows: the common #12 wire used for branch circuits is rated to carry 15 amps; #6 conductors used for feeders and for electric range circuits are rated at 55 amps or 65 amps, depending upon the type of insulation used; #1 conductors can carry 110 amps and are used for 100-amp service entrances for residences. Each size of conductor has a particular current rating for each type of insulation used on the conductor.

Conductor Insulation

Insulation on conductors varies in types and thicknesses for different applications. With few exceptions, wires and cables used in interior electrical systems are insulated. Desirable characteristics of conductor insulation are: high dielectric strength; temperature resistance; flexibility; mechanical strength; moisture resistance; and resistance to arcing and ionization. Of course, no single type of insulation has

1 MIL = $\frac{1}{1000}$ INCH

DIA. = 1 MIL

CIRCULAR MILS = MILS^2

AREA = 1 CM

MCM = 1000 CIRCULAR MILS

- For convenience, wire diameters are expressed in mils rather than in fractions of an inch, where a mil is 1/1000 of an inch.

- Cross-section area of a wire is expressed in circular mils instead of fractions of a square inch, where a circular mil is the area of a circle 1 mil in diameter.

- It can be proved that circular mils are mathematically equal to the square of the diameter in mils.

- The abbreviation CM is used for "circular mils," MCM for "1000 circular mils". Thus 500 MCM stands for 500,000 circular mils cross-section area.

all of the required characteristics for all applications. As a result, a number of different insulations are used to provide best characteristics for various conditions of service.

A LTHOUGH the many types of insulated conductors covered in the National Electrical Code are used for modern power and light circuiting, only about three of the types have received steady, heavy demand for general purpose branch circuit wiring and secondary feeder use over recent past years.

There has been a large measure of standardization in the manufacture of wires for general use. In the way of rubber-insulated wire, Type RHW may be used in both dry and wet locations, with an insulation temperature rating of 75 deg C. This gives the wire higher current carrying capacities than Types R or TW: 65 amps vs 55 amps for No. 6; 100 amps vs 80 amps for No. 3; etc.

Although many engineers, contractors and industrial electrical men have shown preference for RHW wire over plastic-insulated wires—judging the

CONDUCTORS pulled into existing raceway may occupy a greater percentage of the raceway's cross-sectional area than new wiring.	RUBBER		PLASTIC		
	RW RHH	R RH RH-RW RHW	THW	T TW	THWN
#12 OUTSIDE DIAMETER (IN.)	0.221	0.188	0.179	0.148	0.122
#12 CROSS-SECTION AREA (SQ. IN.)	.0384	.0278	.0251	.0172	.0117
NUMBER OF #12 CONDUCTORS PERMITTED IN CONDUIT OR TUBING FOR REWIRING — 1/2-IN. CONDUIT	3	5	5	8	12
3/4-IN. CONDUIT	7	9	10	15	23
1-IN. CONDUIT	11	15	17	25	36

The above table demonstrates the advantage of choosing plastic-insulated conductors for rewiring work. Circles show relative sizes of various insulation types; representative dimensions are given for #12 A.W.G. conductors.

The number of #12 conductors which are permitted by the Code to occupy 1/2, 3/4 and 1-inch conduit are shown for the different insulation thicknesses, based on the requirement that they occupy no greater than 50% of the raceway area.

rubber-insulated wire to be more resistant to cuts and nicks during pulling through raceways and into boxes and considering it to be a more rugged wire for rough handling—plastic-insulated wires have grown steadily in popularity and application. Continual development of new and better materials—polyvinylchloride, in the case of TW and THW wires—and improvement in manufacturing processes by which the plastic is extruded onto the bare wire have made plastic wire a major contender for all sorts of branch circuit and feeder wiring.

Type TW wire, the 60 deg C plastic wire, has smaller overall cross-section area than corresponding sizes of Types R or RHW wire. This smaller overall size of TW wires has given them a big advantage over the standard rubber-insulated wires for rewiring of existing conduits because the NE Code will permit filling of such conduits to 50% of the conduit cross-section area and will permit use of the actual cross-section areas of the conductors on rewiring.

For installation of conductors in conduit (rigid conduit, electrical metallic tubing and flexible metal conduit) for new work, the number of conductors of any given size which may be installed in a given size of conduit is shown in Table 1, Chapter 9 of the Code. The number of conductors given for each combination of wire size and conduit size is based on filling the conduit to 31% of its cross-section when 2 conductors are installed, 43% of its cross section when 3 conductors are installed and 40% of its cross-section when 4 or more conductors are installed.

Table 1 sets the maximum numbers of conductors in various sizes of conduit tubing for a wide range of types of insulated wires—R, RH, RH-RW, RHW, T, TW, THW and, the new wire, THWN. It should be noted that this table sets the same number of permitted wires in a conduit for all of the types of wires. The numbers of wires given in the Table are based on the actual cross-section area of Type RW wire, the largest in overall diameter of the insulated conductors mentioned above. The actual cross-section of the particular wire being used is not a consideration for new conduit wiring.

As a result of Code rules and Table 1, there is no advantage, from the standpoint of how many wires can be gotten in a given size of conduit, in the use of thinwall-insulated plastic wires like TW or THWN when installing new conduit or tubing circuits. For example, a maximum of only 5 No. 12 wires can be installed in a ¾-in. conduit, whether the wires are RHW or THWN, even though the 5 THWN wires would take up a lot less than 40% of the cross-section area of the conduit whereas the 5 RHW wires would fill the 40% area.

For rewiring of existing conduits where it would be impracticable to increase the size of conduit due to structural conditions, the code will permit pulling out the old wires and re-filling the conduit to 50% of its cross-section area when three or more wires are installed in the conduit. And, in the case of rewiring, the actual cross-section area of the conductors can be used in determining how many wires can be installed. For example, if a ¾-in conduit were being rewired to 50% using No. 12 RHW wires, the number permitted would be as follows:

Area of one RHW No. 12 wire = .0384 sq in.
50% of Area of ¾-in. conduit = .27 sq in.

$$\text{No. of Wires permitted} = \frac{.27}{.0384} = 7$$

But if the same conduit were being rewired with No. 12 THWN wires, the number of wires would be as follows:

Area of one THWN No. 12 wire = .0117 sq in.

$$\text{No. of wires permitted} = \frac{.27}{.0117} = 23$$

As can be seen, over three times as many THWN wires can be installed in the ¾-in conduit on rewiring, although only 5 of either RHW or THWN wires can be installed for new work.

For new conduit circuits, when the wires to be installed in the conduit are not all of the same size, Tables

---

CONDUIT AND TUBING DIMENSIONS

FROM NATIONAL ELECTRICAL CODE TABLE 4, CHAPTER 9

TRADE SIZE	INTERNAL DIAMETER (INCHES)	CROSS-SECTION AREA (SQ IN)
½	0.622	0.30
¾	0.824	0.53
1	1.049	0.86
1 ¼	1.380	1.50
1 ½	1.610	2.04
2	2.067	3.36
2 ½	2.469	4.79
3	3.068	7.38
3 ½	3.548	9.90
4	4.026	12.72
5	5.047	20.00
6	6.065	28.89

SAMPLE CALCULATIONS:

A building's branch-circuit wiring is to be replaced and expanded by pulling new conductors into the existing ¾-inch conduit. What is the maximum number of No. 12 conductors that may be used?

For rewiring work, the Code permits the conductors to occupy as much as 50% of the internal cross-section area of the conduit. From the accompanying table, the area of ¾-in. conduit is 0.53 sq. in.

50% OF 0.53 = 0.27 SQ IN.

USING TYPE RW (Area = .0384 sq in):

0.27 ÷ .0384 = 7 CONDUCTORS

USING TYPE RH (Area = .0278 sq. in.):

0.27 ÷ .0278 = 9 CONDUCTORS

USING TYPE THW (Area = .0251 sq. in.):

0.27 ÷ .0251 = 10 CONDUCTORS

USING TYPE TW (Area = .0172 sq. in.):

0.27 ÷ .0172 = 15 CONDUCTORS

USING TYPE THWN (Area = .0117 sq. in.):

0.27 ÷ .0117 = 23 CONDUCTORS

CORRECTION FACTORS *derating the*

normal safe current-carrying capacities of conductors must be applied when the number of conductors or ambient temperature exceeds certain limits:

CONDUIT   CONDUCTORS

I TO 3 CONDUCTORS IN RACEWAY OR CABLE AND OPERATING IN A ROOM TEMPERATURE OF 30C (86F) REQUIRE NO CORRECTION. THEY MAY CARRY THE CURRENTS GIVEN IN CODE TABLES 310-12 AND 310-14.

MORE THAN 3 CONDUCTORS IN RACEWAY OR CABLE REQUIRE A REDUCTION IN CARRYING CAPACITY AS SHOWN IN THE TABLE AT RIGHT.

CONDUCTORS IN RACEWAY OR CABLE OPERATING IN A ROOM TEMPERATURE IN EXCESS OF 30C (86F) REQUIRE A REDUCTION IN CARRYING CAPACITY AS SHOWN IN TABLE.

In each case, the correction factor given is to be multiplied by the normal allowable carrying capacity as given in Tables 310-12 and 310-14 to obtain the reduced allowable current-carrying capacity. These factors help maintain proper insulation temperatures.

CORRECTION FACTORS
FOR MORE THAN 3 CONDUCTORS IN RACEWAY OR CABLE

NUMBER OF CONDUCTORS	CORRECTION FACTOR
I TO 3	NONE
4 TO 6	0.8
7 TO 24	0.7
25 TO 42	0.6
43 & ABOVE	0.5

CORRECTION FACTORS
FOR TEMPERATURES OVER 86F FROM N.E.C. TABLE 310-12

TEMP. C.	TEMP. F.	TYPES R, T, TW	TYPES RH, THW, THWN
40	104	.82	.88
45	113	.71	.82
50	122	.58	.75
55	131	.41	.67
60	140	–	.58
70	158	–	.35

AREAS OF RUBBER-COVERED AND THERMOPLASTIC-COVERED CONDUCTORS
(SQUARE INCHES)
FROM NATIONAL ELECTRICAL CODE, TABLE 5, CHAPTER 9

SIZE A.W.G.	RW RHH*	R RH RHW RH-RW	THW	T TW	THWN
14	.0327	.0230	.0206	.0135	.0087
12	.0384	.0278	.0251	.0172	.0117
10	.0460	.0460	.0311	.0224	.0184
8	.0760	.0760	.0526	.0408	.0317
6	.1238	.1238	.0819	.0819	.0519
4	.1605	.1605	.1087	.1087	.0845
3	.1817	.1817	.1263	.1263	.0995
2	.2067	.2067	.1473	.1473	.1182
1	.2715	.2715	.2027	.2027	.1590
0	.3107	.3107	.2367	.2367	.1893
00	.3578	.3578	.2781	.2781	.2265
000	.4151	.4151	.3288	.3288	.2715
0000	.4840	.4840	.3904	.3904	.3278

* *These dimensions are to be used in computing sizes of conduit for new work.*

EXAMPLE: *If the allowable current-carrying capacity of No. 12 RH under normal conditions is 20 amps, what current may 9 such conductors in one conduit carry?*

From the above table, 7 to 24 conductors in the same conduit require that a correction factor of 0.7 be applied:

0.7 x 20 = **14 AMPS** MAX.

EXAMPLE: *A conduit carrying 3 No. 12 RH conductors passes along a foundry ceiling where the temperature is 125F. What is the maximum current the conductors may carry?*

An ambient temperature of 125F calls for a correction factor of 0.75.

0.75 x 20 = **15 AMPS** MAX.

3, 4 and 5, Chapter 9 of the NE Code will give the maximum number of conductors which can be used in any size of conduit. These tables should be thoroughly familiar to anyone concerned with laying out conduit circuits. Table 3 indicates the allowable fill for conduit under various conditions of use. Table 4 gives the usable area in square inches of various sizes of conduit. And Table 5 gives the cross-section areas of all sizes of Type RW insulated conductors, which must be used regardless of the actual type of wire to be installed.

For rewiring existing conduits, Tables 3, 4A and 5 will give the maximum number of conductors permitted in any size of raceway—whether the conductors are all the same size or are of different sizes.

Current Derating

The Tables given in the Code for Current-Carrying Capacities of conductors in raceway or cable (Tables 310-12 and 310-14) indicate the maximum, continuous, allowable current-carrying capacities for the full range of sizes of the variety of types of insulated conductors when, (A) not more than three conductors are installed in a raceway or cable, and (B) the ambient temperature in which the system is installed is not in excess of 30 deg C (86 deg F).

Where the number of conductors in a raceway or cable exceeds three, the allowable current-carrying capacity of each conductor shall be reduced in accordance with the following Table:

4 to 6 conductors	80%
7 to 24 conductors	70%
25 to 42 conductors	60%
43 and above	50%

The percentages are applied against the value of current given in Table 310-12 or 310-14. For example, if there were 16 Type TW number 8 copper wires in a 2-inch conduit, the allowable current rating of each would be 70% of the 40-amp rating given in Table 310-12 for No. 8 TW, or 28 amps. This indicates the maximum load which can be put on the wires even though they may be protected at the next higher standard value of overcurrent device, 30 amps (for either fuse or circuit breaker).

The above derating of conductor current-carrying capacities applies to rigid metal conduit, electrical metallic tubing, and flexible metal conduit, surface metal raceway, under-floor raceways, headers for cellular metal floor raceways, and headers for cellular concrete floor raceways. In general, all of these raceways are filled to 40% of their cross-section areas. The deratings do not apply to wireways and auxiliary gutters, which must not contain more than 30 conductors (except signal and control wires) at any cross section and these conductors must not occupy more than 20% of the interior cross-section area of the wireway or gutter.

Of course, the derating of conductors when a large number are installed in a raceway is required to keep the heat rise at safe limits within the conduit, to prevent reaching a temperature that would melt or otherwise damage the insulation on the wires. The fact that a large number of wires in a single conduit causes many to be bunched between other wires with very little ventilation or opportunity to dissipate the heat due to current flow (I^2R) makes derating important for proper life of the wires.

As indicated previously, ambient temperature is another consideration which may call for derating of wires below the current values given in Tables 310-12 through 310-15—the tables covering conductors in raceway or in cables or conductors run in free air. In any case, the given current rating is based on a maximum ambient temperature of 30 deg C.

The Tables say, in effect, that each size and type of wire can carry so much current continuously in the given ambient temperature without reaching a temperature level which would be harmful to the insulation on the wire.

For example, a No. 3 Type RHW wire can carry 100 amps continuously when not more than three such wires are installed in a conduit in an ambient temperature not in excess of 30 deg C. Under such conditions, the ambient heat and the I²R heat produced by current flow combine to produce a temperature at the surface of the conductor metal which just comes up to the 75 deg C rating of the RHW insulation. Any more current would push the temperature over the 75 deg C limit.

The way to compensate for increase in ambient over the 30 deg C level is to reduce the allowable current rating of the wire. And this is done by applying the reduction factors given for various temperatures below each of the Tables in the Code. It should be noted that more than three conductors in a raceway or cable also produces excessive heat input even with the ambient level not in excess of 30 deg C. The previously described deratings for more than three wires in a raceway prevents such excessive temperature.

When more than three conductors are installed in a raceway in an ambient above 30 deg C, both deratings have to be made to reduce the allowable current capacity of the conductors for both heat inputs.

THWN Wire

The newest general-purpose building wire is Type THWN, which is covered for the first time in the 1962 NE Code. This is wire with a thermo-plastic insulation (polyvinylchloride) and a tight-fitting outer jacket of transparent nylon. It is a moisture-and-heat-resistant wire, rated for 75 deg C and with the same current carrying capacities as corresponding sizes of Type RHW wire. This wire is suitable for use in dry or wet locations and may be considered as a plastic equivalent of RHW wire.

The overall cross-section area of THWN wire is the smallest of all building wires. This is due to the high insulating and protective characteristics of the plastic insulation and the tough nylon jacket. As noted previously, a No. 12 RHW wire has a cross-section area over three times that of a No. 12 THWN wire. Even the cross-section area of a No. 12 T or TW wire (.0172 sq. in.) is almost 50% larger than the cross-section area of a No. 12 THWN wire (.0117 sq. in.).

As noted before, although the extremely small size of THWN wire does not permit more such wires in a given size of conduit for new work, there is very great advantage in the use of such wire for rewiring. And there is some possibility that future codes may recognize the use of greater numbers of the small cross-section wires in conduit even for new work.

The nylon jacket on THWN wire gives it excellent resistance to abrasion and cutting on sharp box edges during pulling-in, eliminating the objection that many engineers and contractors have had toward plastic-insulated wires. It is claimed that the jacket toughness of THWN wire is even better than that of any other general-use building wire. And the nylon jacket is highly resistant to gasoline, oils, chemicals and other corrosive atmospheres—although it is not specifically made for gasoline locations.

Other characteristics of THWN wire are as follows:
Extremely low coefficient of friction permits longer pulls through more bends and offsets than other types of wire. The very low friction between the nylon jacket and the conduit wall can greatly reduce the man-hours of installation labor, sometimes even eliminating the use of power pulling winches which would be necessary with the same number of other types of wire under identical conditions.
High- and low-temperature workability of the nylon jacket makes the overall wire flexible enough for installation even at —10 deg C, without cracking of the wire covering due to repeated flexures. Temperature applications up to 105 deg C are also acceptable.

INSULATED conductors are generally covered with a protective mechanical sheathing to prevent damage to the insulation and to inhibit deterioration of the insulation due to job conditions or atmospheres. Such protection on conductors varies widely in types and application. Each type of protective conductor covering has its own characteristics and is suited to certain classes of installation:

Fibrous braids of many types are used as conductor coverings. On a conductor, a braid covering is tightly woven over the insulation and is usually saturated with a compound to give added resistance to heat, moisture and corrosive atmospheres. Fibrous braids are commonly used on low-voltage conductors and cables. Fibrous tape coverings are used for inner coverings in protective coverings for cables.

Rubber and synthetic jackets are also common conductor and cable coverings. The compounds used for these coverings are known by various trade names, depending upon the manufacturer. They are tough and resistant to the many mechanical and atmospheric conditions against which wires and cables need protection.

Lead sheathing is an old standard cable covering for use in wet locations and for many underground applications. Where corrosion of the lead is a possibility, one of the other protective coverings is used over the lead jacket. There are several types of lead covered cables suited to particular applications.

Armor cable coverings are metallic wraps which provide a strong and flexible type of protection. One type of armor covering consists of two steel tapes wrapped around the cable, spiralled in such a way that each tape covers the space between successive turns of the other tape. Another type is interlocked armor cable, in which a single interlocking metal tape is spiralled along the length of the cable.

The tape has a rounded cross-section which adds strength and flexibility to the covering. Adjacent turns on the tape are tightly locked to each other, providing a continuous interlocked armor for the entire length of the cable.

Wire armor is another type of metallic covering consisting of a layer of round metal wires spiralled tightly around the cable. Galvanized steel wires are usually used for the wire armor, although non-ferrous wires are also used. Basket-weave armor is a cable covering which consists of a braid of metal wires woven over the cable.

Types of Cables

Cables used in electrical systems and equipment may be classified according to their uses as follows:

General purpose building wires and cables are generally considered as a single group of conductors and assemblies. These are the rubber-, thermoplastic-, varnished-cambric-, paper- or asbestos-insulated conductors covered singly or in multi-conductor assemblies. Coverings may be fibrous or rubber or synthetic jacketings for general applications. Asbestos braid is used for the protective covering when high resistance to flame is required. Lead sheath cable is used for underground duct systems and in wet locations. Rubber or synthetic sheaths like neoprene are used for various hardduty applications. Thermoplastic and latex-insulated conductors do not have outer coverings and are widely used in interior wiring systems and equipment wiring. General-purpose power cables are made for both low-voltage uses (under 600 volts) and for high-voltage applications (over 600 volts).

Type UF cable has a non-fibrous covering which is highly resistant to moisture, heat, abrasion and corroding agents. This cable is used for underground branch circuits and feeders, directly buried in the ground, either single- or multi-conductor.

Non-metallic sheathed cable is used for wiring of homes, farm buildings and small commercial occupancies. Type NM cable has a fibrous covering which gives the assembly high mechanical strength. Type NMC has a non-fibrous covering which has all the strength of NM but is also corrosion resistant. Type NMC is used in wiring barns, where a corrosive atmosphere exists. Single- or multi-conductor cables of both types are available.

Armored building cables are available in several types. Type AC cable has rubber- or thermoplastic-insulated conductors protected by a steel armor. The cable with thermoplastic-insulated conductors is classed Type ACT. A common name for such cable is "BX" (actually a manufacturer's trade name for Type AC). For wet locations, armored cable with an inner sheath of lead is available.

Interlocked armor cable is "king-sized" armored cable of the type just described. It is made in sizes larger than regular armored cable (which is made up to #1) and is finding increasing use in feeder applications at low and high voltages.

Aerial cables are made in several types, some bare and other insulated and/or covered. Power transmission lines are usually bare cables. Weatherproof cable is used for distribution and low-voltage transmission. Covered rubber-insulated conductors make up "tree wire" used for aerial lines where trees interfere with the lines. A number of multi-conductor aerial cables are also used for outdoor lines, usually supported by a steel messenger cable to which the power cable is tied. Self-supporting aerial cable is an assembly of heavily-insulated power cable which can be installed without the use of messenger cable for support, or is a group of power cables made up as a complete assembly with a supporting messenger.

Service cables are assemblies of insulated conductors (although the neutral conductor is frequently bare). Type SD is service drop cable for connecting from the utility pole line to the premises to which power is delivered. Type SE is service entrance cable, which connects to a service drop and carries power into the service equipment of a building. Type ASE is service cable with metal interlocking armor as a protective covering, called "armored service entrance" cable. Type USE is underground service entrance for installation in conduit or directly buried in the earth.

MANY special cable assemblies are made with mechanical and electrical characteristics suited to particular applications. Construction and application data are as follows:

Control cables are assemblies of insulated wires used for various control circuits to electrical machines and equipment. Wires used in such cable may be solid or stranded conductors of small diameter to handle the normally low currents used for control. Each conductor may be rubber-insulated, thermoplastic-insulated, varnished cambric-insulated or asbestos-insulated for high temperature installations. The individual conductors are generally covered with a protective cotton braid and are color coded for easy identification of the different circuits. A typical control cable might contain one conductor or more than thirty conductors. The complete assembly of individual conductors has an overall covering or jacket forming the cable. Non-metallic cable covering may be a rubber overall jacket on the conductor assembly or may be a treated braid over rubber tape. Metallic covering may be lead sheath or metallic armor.

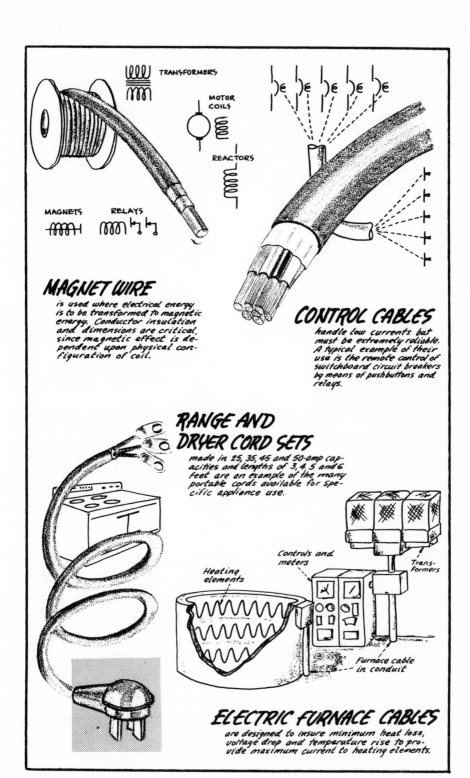

TRANSFORMERS

MOTOR COILS

REACTORS

MAGNETS **RELAYS**

MAGNET WIRE

is used where electrical energy is to be transformed to magnetic energy. Conductor insulation and dimensions are critical, since magnetic effect is dependent upon physical configuration of coil.

CONTROL CABLES

handle low currents but must be extremely reliable. A typical example of their use is the remote control of switchboard circuit breakers by means of pushbuttons and relays.

RANGE AND DRYER CORD SETS

made in 25, 35, 45 and 50-amp capacities and lengths of 3, 4, 5 and 6 feet are an example of the many portable cords available for specific appliance use.

Controls and meters

Transformers

Heating elements

Furnace cable in conduit

ELECTRIC FURNACE CABLES

are designed to insure minimum heat loss, voltage drop and temperature rise to provide maximum current to heating elements.

Type ALS Cable

A new cable assembly covered for the first time in the 1962 Code is Aluminum Sheathed Cable — designated Type ALS. This cable consists of insulated wires in an impervious, continuous, closely fitting, seamless tube of aluminum. This cable is recognized for both exposed and concealed work in dry or wet locations, with approved fittings.

Type ALS cable is an alternative wiring method to rigid conduit, EMT, flexible metal conduit and other cable types. It has the advantage that no couplings are required for long runs, the jacket can be easily removed with a pen knife or a tubing cutter and bends and other changes in direction can usually be made by hand, especially with the smaller sizes. The cable must be secured at intervals not greater than 6 feet by means of staples, straps, hangers or similar fittings.

Although not approved for hazardous locations, ALS cable can be used for the full range of commercial and industrial applications, including use in theatres and other places of public assembly as permitted by Code Article 520.

Control cables are made for use in wet or dry locations and in atmospheres of chemicals and corrosive fumes. In commercial and industrial applications, these cables are used to wire motors to their controllers and to remote relays or control devices, to wire signal lighting circuits, to wire traffic lighting circuits, to wire relay circuits for light and power control, to wire metering circuits and to wire remote-operated circuit breakers.

Mine cables are made in many types for use in all mining operations. Wire and cable constructions for mine use are suited to particular voltages under various conditions of use in underground and open-pit mines. This includes high-voltage electrical distribution, low-voltage (under 600 volts) distribution and branch circuits to outdoor and indoor loads.

Typical special wires and cables for the mining industry are: borehole or mine-shaft cables which are rubber or varnished cambric insulated, with or without a lead sheath, constructed to stand the strain of long vertically-suspended lengths; bare trolley wires for powering moving mining machines; cords and cables for use on mining machines — shuttle cars, operating drills, cutting machines, locomotives, hand drills; power cables for feeders.

Submarine cables are cables constructed for use in electrical circuits under water. They are used in rivers, harbors, lakes, etc. Such cables are covered with either a highly water resistant compound or with an impervious sheath such as lead, providing strong mechanical protection against underwater obstructions, rocks, ship anchors, buffeting due to tides and similar rough treatment.

Submarine cables are used for low or high voltage circuits in electrical transmission or distribution systems. Design of submarine cable installations requires special engineering for the particular conditions.

Network cables are special cables used for low voltage electrical power networks which provide high reliability in the transfer of power from one or more supplies to load circuits. This cable is used by utility companies in their networks, installed underground in ducts or in overhead lines. It is also used in industrial plant network distribution systems. Such cables are generally single-conductor cables — of paper and lead, rubber and lead or rubber or neoprene construction, depending upon operating conditions.

Station cables are used for hydroelectric, steam and diesel power stations, for indoor and outdoor substations, for generator and transformer leads, for bus connections and other power cable applications. A wide range of constructions is available.

Vertical riser cable is used for vertical cable runs in shafts, between a transformer vault and an aerial line, connecting underground systems to overhead lines and similar applications. The cable has construction characteristics suited to the widely varying weather and atmospheric conditions it will encounter and to the conditions of strain in installation.

Portable power cables are used for supplying power to portable power equipment—dredgers, welders, compressors, electric cranes, electric shovels, and similar equipment. Both low and high voltage types are available with suitable protective coverings.

Gas-tube-sign and ignition cable is another special type of cable, used for transformer circuits to gaseous discharge lamps and for oil-burner ignition circuits. Such cable is designated by Underwriters as follows: GTO-5 for use up to 5,000 volts; GTO-10 for use up to 10,000 volts; and GTO-15 for use up to 15,000 volts.

Airport lighting cables are special cables which meet the specifications of the Civil Aeronautics Administration for use in airports. Such cables are single- and multi-conductor types— type R or RH conductor insulation— with an overall neoprene jacket, suited to wide application for direct-burial or in-duct underground circuits to obstruction lights, runway lights, rotating beacons, etc.

Electric furnace cables are very flexible high-current, low-voltage cables used for connecting to the electrodes in electric arc furnaces and resistance furnaces. Asbestos and varnished-cambric insulation is used on such cables.

Series street-lighting cables have rubber, thermoplastic, paper or varnished-cambric insulated conductors, with an outer jacket of lead or neoprene. These are generally single-conductor cables in ratings up to 15,000 volts.

Magnet wires are single-conductor insulated wires used for winding coils for various types of electrical equipment. Round magnet wire is made in sizes from No. 46 to No. 4/0. Square and rectangular magnet wires are also made. Typical magnet wire insulations include: enamel, cotton, paper, asbestos, silk and fiber glass. The different insulations are suited to various applications. Typical coils wound of magnet wire are: relay coils, motor coils, transformer coils, reactor coils, ballast coils, contactor coils and magnet coils.

Electronic wires and cables cover a wide range of single-conductor and multi-conductor cables, rubber or thermoplastic insulated, and rubber or thermoplastic insulated wires. These wires and cables are used in radios, televisions, electronic devices, for hookup of sound systems and signal systems and for control circuits.

Protective Devices

ELECTRICAL wires, cables, devices, equipment and apparatus are rated according to the normal value of voltage which can be applied to them and the current which they can safely carry. If other than the safe value of voltage were applied to a conductor or piece of equipment, or if higher than rated current were to flow, the conductor or piece of equipment would be seriously damaged or completely destroyed. In addition, a general condition of hazard to personnel would result. To protect conductors and equipment against abnormal operating conditions and their consequences, a special group of devices are used in electrical systems. These devices include fuses, circuit breakers and relays and are commonly called "protective devices."

Fuses

A fuse is the simplest, best known and most widely used device for opening an electrical circuit when excessive current flows due to overload or short circuit. Basically, a fuse is a series circuit device which will conduct the current for which the circuit (and the fuse) is rated; but it has an internal element, called a "link," which will melt and open the circuit when excessive current flows.

There are many types and sizes of fuses used to provide overload protection for both circuits and apparatus. Enclosed fuses range in capacity up to 600 amps at 250 and 600 volts. Fuses for high voltage applications (above 600 volts) are available with current capacities up to 400 amps and voltage ratings to 132,000 volts. Fuses will carry their maximum rated current continuously but will blow in from 1 to 5 minutes if current exceeds a 15 per cent rise over rated capacity.

• **Plug Fuses**—These are screw-in type fuses. The Edison-base plug fuse is familiar to everyone from its wide application as a branch circuit protector in residential electrical systems.

An Edison-base plug fuse consists of a wire or strip of fusible metal in a small porcelain case with a mica disk set in the top making the fuse link visible. The screw base corresponds to the base on a standard medium-base incandescent lamp. These fuses are confined to use on circuits rated at a maximum of 125 volts. They are made in various current ratings: 3, 6, 10, 12, 15, 20, 25 and 30 amps. Those fuses rated at 15 amps or less have an hexagonal opening in the cap, through which the mica shows, or they may have an hexagonal recess or projection in the top, or the top itself may be hexagonal in form. Those fuses rated

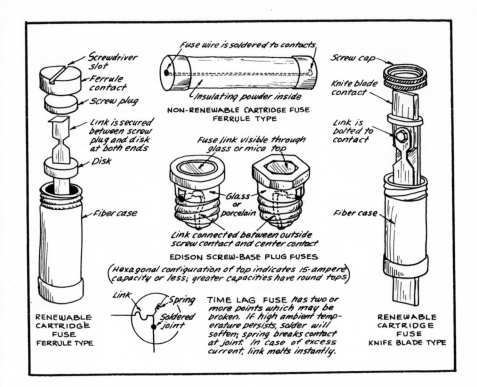

Screwdriver slot
Ferrule contact
Screw plug
Link is secured between screw plug and disk at both ends
Disk
Fiber case

Fuse wire is soldered to contacts
Insulating powder inside
NON-RENEWABLE CARTRIDGE FUSE FERRULE TYPE

Screw cap
Knife blade contact
Link is bolted to contact
Fiber case

Fuse link visible through glass or mica top
Glass or porcelain
Link connected between outside screw contact and center contact
EDISON SCREW-BASE PLUG FUSES
(Hexagonal configuration of top indicates 15-ampere capacity or less; greater capacities have round tops)

Link
Spring
Soldered joint
TIME LAG FUSE *has two or more points which may be broken. If high ambient temperature persists, solder will soften; spring breaks contact at joint. In case of excess current, link melts instantly.*

RENEWABLE CARTRIDGE FUSE FERRULE TYPE

RENEWABLE CARTRIDGE FUSE KNIFE BLADE TYPE

from 16 to 30 amps have a round window in the top. The purpose of this is to make the lower rated fuses easily distinguishable from the higher rated fuses to minimize accidental interchange of the fuses.

Plug fuses and fuse holders classified as Type S are similar in general construction to the Edison-base plug fuse and its holder. Type S fuses, however, are tamper-resisting, i.e., fuses of the 0 to 15 amp sizes cannot be interchanged with fuses of the 16 to 30 amp sizes. This arrangement eliminates the possible hazards of overfused (too large a fuse) circuits, in which the current may exceed the rating of the circuit conductor and load devices but does not exceed the rating of the fuse.

The use of tamper-resisting plug fuses is widely recommended. These fuses are available with the same ratings as Edison-base plug fuses. The tamper-resisting characteristic is ac-

complished by using a special fuse construction and adapter. The adapter is first screwed on the fuse base. Then the combination is screwed into a regular Edison-base cutout or holder. If the fuse should then be removed from the holder, the adapter stays in the holder, and only the same size Type S fuse can be replaced in the holder. An ordinary plug fuse cannot be inserted in the holder, and a Type S fuse of the wrong group rating cannot be inserted.

Time-delay plug fuses are fuses which allow the abnormal circuit operating condition to persist for a longer time before blowing than ordinary fuses. Also called "time-lag" fuses, they are available in regular Edison-base type plug and tamper-resisting screw base up to 30 amps. These are one-time (non-renewable) fuses. The Edison-base type and tamper-resisting base type are identical except for the base construction. Again the tamper-

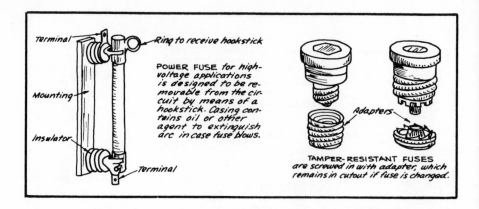

Terminal

Ring to receive hookstick

POWER FUSE for high-voltage applications is designed to be re-movable from the cir-cuit by means of a hookstick. Casing con-tains oil or other agent to extinguish arc in case fuse blows.

Mounting

Insulator

Terminal

Adapters

TAMPER-RESISTANT FUSES are screwed in with adapter, which remains in cutout if fuse is changed.

resisting type needs an adapter before it can be screwed into an Edison plug-fuse receptacle. Adapters are made in three common sizes: a 15-amp adapter which will take any size fuse up to 15 amps; a 20-amp adapter for a 20-amp fuse; and a 30-amp adapter for a 20-, 25- or 30-amp fuse.

The time-delay type fuse is widely used on motor circuits in which the starting current inrush is much higher than the continuous current. The time-lag fuse will not open on the short duration of high starting current. If, however, the high current should persist past the delay time of the fuse, it will blow just as if a short circuit or heavy overload current had developed.

• **Cartridge Fuses**—A typical car-tridge fuse consists of a tube of stiff, treated paper, fiber or some similar material, with the fuse element within the tube. A contact piece of some type is mounted on each end of the tube. The fuse element is connected between these metallic contact pieces inside the tube:

In a nonrenewable (one-time) car-tridge fuse, the fuse element consists of a fuse wire stretched through the tube in an insulating compound be-tween copper ferrule terminals on the ends of the tube. When the fuse blows, the compound quelches any possible arc.

Renewable cartridge fuses contain fuse links which can be replaced when blown. This type of cartridge fuse can be readily taken apart for simple and easy replacement of a blown link. Renewable fuses are made in types with copper ferrule terminals at the tube ends and with knife-blade con-tacts.

Cartridge fuses have ratings and dimensions as follows: 0 to 250 volts, 0 to 60 amps—2 inches and 3 inches long, ¾-inch diameter; 0 to 250 volts, 61 to 600 amps—up to 11 inches long, 2½ inches in diameter; 251 to 600 volts, 0 to 600 amps—5 inches to 14 inches long, 3 inches in diameter.

OVER RECENT YEARS, growth in the extent and complexity of electrical systems for power and lighting has stimulated development of a wide variety of cartridge fuses to meet widely varying operating de-mands.

Developments have included: in-crease in the amount of short-circuit current which fuses can safely inter-rupt; current-limiting action to snuff out a fault so fast that the current never reaches its maximum level; time-delay characteristics in high-capacity fuses; and rejection features in fuse terminals and fuse holders, to prevent replacement of current-limiting fuses by non-current-limiting fuses.

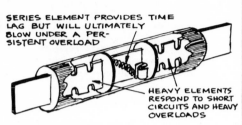

SERIES ELEMENT PROVIDES TIME LAG BUT WILL ULTIMATELY BLOW UNDER A PERSISTENT OVERLOAD

HEAVY ELEMENTS RESPOND TO SHORT CIRCUITS AND HEAVY OVERLOADS

COMMON FUSES
AND THEIR USES

- **PLUG FUSES**
  - LIGHTING & APPLIANCE BRANCH CIRCUITS
- **DUAL-ELEMENT PLUG FUSES**
  STANDARD SCREW BASE & TYPE "S"
  - ANY TYPE BRANCH CIRCUIT
  - SMALL MOTOR PROTECTION
- **ONE-TIME CARTRIDGE FUSES**
  - MAINS AND FEEDERS; CIRCUITS OF INFREQUENT TROUBLE
- **RENEWABLE CARTRIDGE FUSES**
  - MAINS AND FEEDERS; CIRCUITS WHERE FAULTS ARE FREQUENT
- **DUAL-ELEMENT CARTRIDGE FUSES**
  - MAINS AND FEEDERS; MOTOR OR APPLIANCE BRANCH CIRCUITS; MOTOR OVERLOAD PROTECTION
- **SILVER-SAND FUSES**
  HIGH-INTERRUPTING TYPE
  - MAINS AND FEEDERS REQUIRING HIGH INTERRUPTING CAPACITY

  CURRENT-LIMITING TYPE
  - MAINS AND FEEDERS REQUIRING FAULT CURRENT TO BE LIMITED

DUAL-ELEMENT cartridge fuses are used in any type of feeder circuits and branch circuits and for motor protection. A typical dual-element fuse may have a time delay of 12.8 seconds at 5 times normal load, whereas the same size single-element fuse would have a delay of only 0.7 second. A blown fuse in a motor circuit could indicate such abnormal operating conditions as an overloaded motor, a tight or worn bearing, a tight belt, a dirty commutator, or incorrectly placed brushes.

Application of fuses demands an understanding of their rating. Fuses are tested at a constant current on a 60-cycle circuit. They are tested in the open, mounted in a horizontal position on a bench or board—in fuseholders for fuses up to 600 amps and bolted to busbars for fuses of higher current rating. In this test, a fuse must be capable of carrying 110% of rated current continuously, without any damage to the assembly.

Because fuses are rated in open air, consideration must be given to effective derating of fuses when they are enclosed under conditions of heat accumulation. It is common to derate fuses about 10% and take the rating of 100% of rated current value when fuses are enclosed by switch or panel enclosures.

Consideration of heat accumulation in the case of enclosed fuses is made in NEMA standards on safety switches. To assure continuous opera-

tion of loads—minimizing nuisance fuse blowing—NEMA requires that a fused enclosed switch be marked, as part of the electrical rating, "Continuous Load Current Not To Exceed 80% Of The Rating Of Fuses Employed In Other Than Motor Circuits." Thus fusible switches must be rated to accommodate fuses which are at least 25% higher in rating than load currents which will be flowing for several hours or longer.

This NEMA requirement is often misunderstood as being a rule of the NE Code that switches have to be derated to 80% of their nameplate current rating. The NE Code has no such rules. And it should be noted that the NEMA rule applies only to fused switches.

Cartridge fuses can be broken down as follows:

1. Single-element fuses have a current-responsive element of a single fusing characteristic. These are the so-

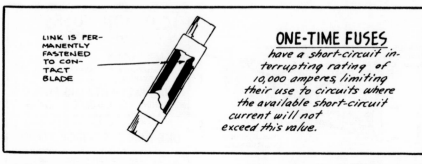

ONE-TIME FUSES
have a short-circuit in-
terrupting rating of
10,000 amperes, limiting
their use to circuits where
the available short-circuit
current will not
exceed this value.

called "standard NEC (National Electrical Code) fuses" and are available in two forms:

A. Non-renewable fuse, called a "one-time" fuse, cannot be restored for service after it blows and must be replaced. Such fuses are made for either 250 volts or 600 volts, up to 600 amps. An ordinary one-time fuse has comparatively little time delay in its operation. It will carry 110% of its rated current continuously in open air and about 100% of rated current when installed in an enclosure. At currents above 110% rated value, the fuse will blow. It will blow after some time delay on light overloads and will blow quickly on heavy overloads or shorts.

Because of their lack of time delay in operation, one-time fuses are generally susceptible to blowing on harmless overloads and motor starting currents. They are suitable for use in circuits with little chance of operating overloads, such as lighting circuits, where the fuse guards against grounds and shorts. When used on motor circuits, one-time fuses must take full advantage of Code permission to use fuses rated up to three or four times motor full-load running current, in order to withstand the starting inrush current without blowing and opening the circuit.

B. Renewable fuse is one which may be readily restored for service after operation by replacement of the renewal element, the current-responsive single element which provides the protection. This fuse is electrically similar to the one-time fuse and is made in the same current and voltage ranges. But through more detailed design of the fuse link and terminal assembly, renewable fuses are commonly made with longer time delay than standard one-time fuses.

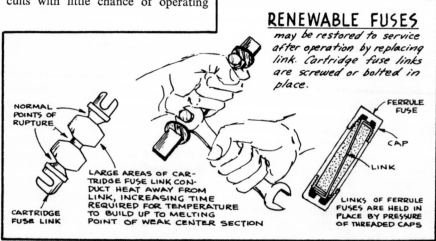

RENEWABLE FUSES
may be restored to service
after operation by replacing
link. Cartridge fuse links
are screwed or bolted in
place.

NORMAL
POINTS OF
RUPTURE

LARGE AREAS OF CAR-
TRIDGE FUSE LINK CON-
DUCT HEAT AWAY FROM
LINK, INCREASING TIME
REQUIRED FOR TEMPERATURE
TO BUILD UP TO MELTING
POINT OF WEAK CENTER SECTION

CARTRIDGE
FUSE LINK

FERRULE
FUSE

CAP

LINK

LINKS OF FERRULE
FUSES ARE HELD IN
PLACE BY PRESSURE
OF THREADED CAPS

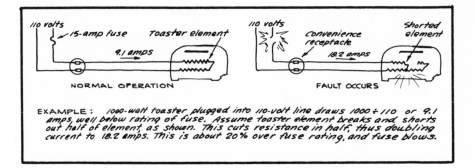

EXAMPLE : 1000-watt toaster plugged into 110-volt line draws 1000 ÷ 110 or 9.1 amps, well below rating of fuse. Assume toaster element breaks and shorts out half of element as shown. This cuts resistance in half, thus doubling current to 18.2 amps. This is about 20% over fuse rating, and fuse blows.

Although renewable fuses cost more at first than one-time fuses, they offer the economy of repeated renewal of the fusible link, which is much less expensive than replacing the entire fuse. And the greater time delay of renewable fuses makes possible the use of a lower rated fuse than a one-time fuse for motor circuits with their high inrush currents. This will often permit the use of a smaller switch than would be necessary to accommodate the required size of one-time fuses.

2. Dual-element fuse has two current-responsive elements of different fusing characteristics in series in a single cartridge. One of the elements provides extended time-delay operation on overloads and the other element will respond only to very high currents, like short circuits, to open the circuit instantly. Such fuses have much greater time delay than single-element, time-delay fuses.

The very long time delay of dual-element fuses enables them to handle much higher motor starting currents, without blowing, than time-delay single-element fuses. Thus they represent a further extension of the application possibilities for reduced-size fusing described for renewable fuses. Instead of fuses rated at 300% of motor current to provide short-circuit protection of the branch circuit without blowing on motor starting current, fuses with current rating equal to the motor full-load current, or only slightly higher—say, 125%, can be used in many installations. Again, in very many cases, this will permit use of smaller switches than would be required for accommodating single-element fuses. And, in many applications, when the rating of the fuses does not exceed 125% of motor current, the same fuses may simultaneously provide the required running overload protection for motor and controller, in addition to providing short-circuit protection.

Although dual-element fuses are particularly applicable to motor branch circuits and feeders, they are also ideally suited to all types of circuits for power loads, mixed lighting and power loads and lighting loads alone. They are made with very high interrupting capabilities—up to 100,-000 amps. And the low resistance construction of such fuses offers much cooler operation, providing higher loading of fuses in switch or panel enclosures without heat damage to switches or other equipment and minimizing nuisance blowing of fuses due to accumulated ambient heat.

OVER RECENT YEARS, tremendous growth in the generating and transmission capacity of electric supply systems has brought very serious potential hazard to life and property from the destructive thermal and magnetic forces produced by the extremely high fault currents (grounds and shorts) available in modern circuits.

Effective protection against such high fault currents and their destructive potential can be provided by the use of current-limiting fuses.

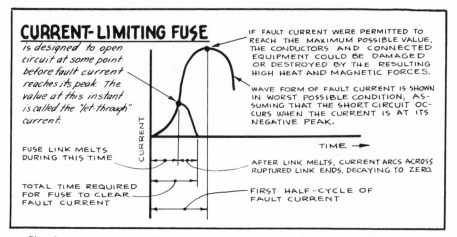

CURRENT-LIMITING FUSE

is designed to open circuit at some point before fault current reaches its peak. The value at this instant is called the "let-through" current.

IF FAULT CURRENT WERE PERMITTED TO REACH THE MAXIMUM POSSIBLE VALUE, THE CONDUCTORS AND CONNECTED EQUIPMENT COULD BE DAMAGED OR DESTROYED BY THE RESULTING HIGH HEAT AND MAGNETIC FORCES.

WAVE FORM OF FAULT CURRENT IS SHOWN IN WORST POSSIBLE CONDITION, ASSUMING THAT THE SHORT CIRCUIT OCCURS WHEN THE CURRENT IS AT ITS NEGATIVE PEAK.

CURRENT

TIME →

FUSE LINK MELTS DURING THIS TIME

AFTER LINK MELTS, CURRENT ARCS ACROSS RUPTURED LINK ENDS, DECAYING TO ZERO.

TOTAL TIME REQUIRED FOR FUSE TO CLEAR FAULT CURRENT

FIRST HALF-CYCLE OF FAULT CURRENT

Simply stated, a current-limiting fuse is a fuse which safely interrupts all available currents within its interrupting rating and, within its current-limiting range, limits the clearing time at rated voltage to an interval equal to or less than the first major or symmetrical current loop duration and also limits peak let-through current to a value less than the peak current that would be possible if the fuse were replaced with a solid conductor. This means that a current-limiting fuse can put a ceiling on the amount of short-circuit current that can flow in a given circuit, regardless of the amount of current which the supply system could deliver into a short circuit if the fuse were not present to limit the current.

Operation of a current-limiting fuse is somewhat involved and even complex. But, effective application of such fuses depends upon a sure knowledge of their characteristics and advantages.

And the extremely important role current-limiting fuses play in making modern electrical systems safe demands this sure knowledge:

1. Every fuse is designed to interrupt the current flowing through it when the current exceeds the continuous rating of the fuse. As described previously, this means that a fuse will carry 100% of its rated current when enclosed in a panelboard or switch enclosure but will blow at higher currents. On light overloads, the fuse will have some time delay before it blows, depending upon its particular design. On heavy overloads, the fuse will open faster than on light overloads. And on short circuits, the fuse will open even faster. Basically, the higher the current, the faster the fuse will blow and open. This is due to the fact that the fuse is responding to heat energy, which is equal to the heat input to the fuse produced by the current (I^2R)

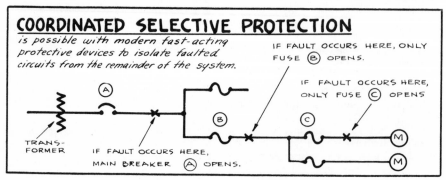

COORDINATED SELECTIVE PROTECTION

is possible with modern fast-acting protective devices to isolate faulted circuits from the remainder of the system.

IF FAULT OCCURS HERE, ONLY FUSE Ⓑ OPENS.

IF FAULT OCCURS HERE, ONLY FUSE Ⓒ OPENS

Ⓐ

Ⓑ

Ⓒ

TRANS-FORMER

IF FAULT OCCURS HERE, MAIN BREAKER Ⓐ OPENS.

Ⓜ

Ⓜ

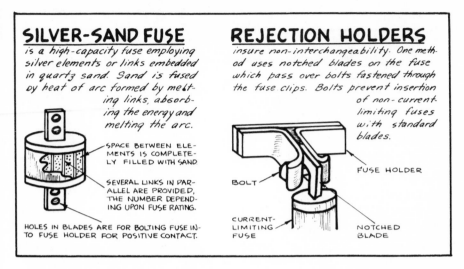

SILVER-SAND FUSE

is a high-capacity fuse employing silver elements or links embedded in quartz sand. Sand is fused by heat of arc formed by melting links, absorbing the energy and melting the arc.

SPACE BETWEEN ELEMENTS IS COMPLETELY FILLED WITH SAND.

SEVERAL LINKS IN PARALLEL ARE PROVIDED, THE NUMBER DEPENDING UPON FUSE RATING.

HOLES IN BLADES ARE FOR BOLTING FUSE INTO FUSE HOLDER FOR POSITIVE CONTACT.

REJECTION HOLDERS

insure non-interchangeability. One method uses notched blades on the fuse which pass over bolts fastened through the fuse clips. Bolts prevent insertion of non-current-limiting fuses with standard blades.

FUSE HOLDER

BOLT

CURRENT-LIMITING FUSE

NOTCHED BLADE

---

times the time that the current flows. As the current increases, I^2R value increases quickly and the time required to produce the energy to melt the fuse link decreases.

2. If a value of 60-cycle short circuit current within the interrupting rating of the fuse causes the fuse to open in less than 1/120th of a second, the current has not gone through one complete half-cycle loop. As a result, the fault current has not reached the rms value which it would reach if just one half cycle were completed.

Starting from zero, an ac wave reaches its rms value when the first half cycle is complete. Each succeeding half cycle does not represent an increase in the rms value but simply represents the same value of rms which will obtain until the circuit is broken. Any fuse, then, that opens before a half cycle of current flows has actually reduced or "limited" the current. But the standard for current-limiting fuses also states that the fuse must limit the peak value of the current it passes before opening to a value less than the peak value normally obtained in the current wave which would flow if the fuse were not there to open the circuit.

3. Any fuse which can produce the above action within its operating range is actually a current-limiting fuse. As

a result, standard single-element and dual-element fuses, which can produce some measure of current limitation, are current-limiting fuses. But the term "current-limiting" is reserved to fuses which are especially designed with a high order of current limitation and high interrupting capacity.

Current-limiting fuses, from NEMA Standards, have an interrupting rating of 200,000 rms amperes. They are made in cartridge type with ferrule terminals up to 60 amps and with knife blade terminals from 70 to 600 amps. In larger sizes, up to 6000 amps, they are made with stud-mounting (bolt-in) terminal blades. Typical fuses are made for use at 250 volts or less. Others are made for 600 volts or less.

Although current-limiting fuses are made in standard ferrule and blade types, they are also made with special ferrules and blades, affording ready insertion in special rejection fuseholders which will not accept standard NEC fuses but still permitting their use in standard fuseholders. This is available to satisfy NE Code Sec. 240-23(b), which says, "Fuseholders for current-limiting fuses shall not permit insertion of fuses which are not current limiting." Such an arrangement provides non-interchangeability to prevent a specific hazard.

When equipment is used in conjunc-

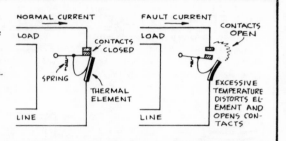

THERMAL BREAKER

depends upon action of heat on bi-metallic element to open contacts. Some breakers incorporate an additional compensating element to counteract effect of ambient temperatures on main thermal element.

NORMAL CURRENT

LOAD

CONTACTS CLOSED

SPRING

THERMAL ELEMENT

LINE

FAULT CURRENT

LOAD

CONTACTS OPEN

EXCESSIVE TEMPERATURE DISTORTS ELEMENT AND OPENS CONTACTS

LINE

tion with current-limiting fuses—with busbars and other parts constructed and braced to withstand only the limited energy let-through by such fuses —replacement of the current limiting fuses by non-current-limiting types could expose the busbars and parts to the much higher levels of destructive energy which non-current-limiting fuses will pass. If, when current-limiting fuses are used, they are used in conjunction with rejection holders, the fuses cannot be replaced by non-current-limiting types.

Current-limiting fuses are especially suited to high capacity service conductors, feeders and subfeeders, where fast operation of the fuse is desired on short-circuit faults, limiting the current to a value which will not damage or destroy equipment—such as switches or busway—which will have to carry the fault current until the fuse opens.

Another type of fuse in wide application is the "high capacity" fuse. This is a one-time fuse for circuits from

600 to 6,000 amps, with some current limitation but not enough to qualify as a true "current-limiting" fuse. Such a fuse has longer time delay than a current-limiting fuse and is well suited to use as main protection supplying a number of feeders which are protected by current-limiting fuses. In such hookup, a fault on a feeder will operate the much faster acting current-limiting fuse and not blow the main fuse. Thus the effect of a fault is confined to the faulted feeder, and all feeders are not knocked out by blowing of the main.

The dual-element, current-limiting fuse is a fuse which combines the time delay on overloads of the dual-element fuse, with fast action and current limitation of the current-limiting fuse. They are suited to use in motor control centers where the time delay permits motor starting but where the current limitation protects against the destructive thermal and magnetic forces of very high short-circuit currents by limiting the current to a lower value.

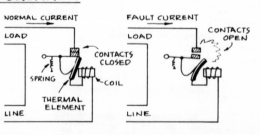

THERMAL-MAGNETIC BREAKER

includes both a thermal element and a magnetic coil. Short-circuit current increases pull of magnetic device sufficiently to open contacts. Persistent overload, while not great enough to operate magnet, heats thermal element and opens circuit.

NORMAL CURRENT

LOAD

CONTACTS CLOSED

SPRING

COIL

THERMAL ELEMENT

LINE

FAULT CURRENT

LOAD

CONTACTS OPEN

LINE

Circuit Breakers

In addition to fuses, circuit breakers are devices which find widespread application as circuit and equipment protective devices. A circuit breaker is simply a switching device which will provide automatic interruption of current flowing through it when the current conditions are abnormal. And it will accomplish this action without damage to itself. A circuit breaker mechanism is set to interrupt the current at a particular overload value, and it can interrupt short circuit currents. A switch can interrupt its load current but not short circuit current.

The automatic circuit-opening action of a circuit breaker may be obtained in several ways: by thermal release, by magnetic action, by some combination of thermal and magnetic or hydraulic principles. Relays are used to obtain opening action in electrically operated breakers; simple release devices are used in manually operated breakers.

Circuit breakers are available in a wide range of sizes for many applications. Small breakers for 15, 20, 30 and 50-amp branch circuits are widely used today in service entrance equip-

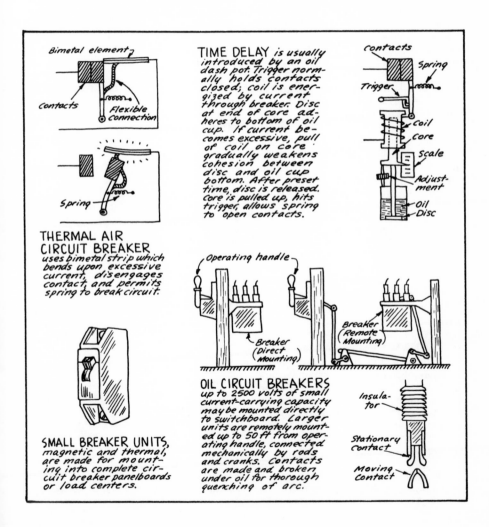

Bimetal element

Contacts

Flexible Connection

Spring

THERMAL AIR CIRCUIT BREAKER
uses bimetal strip which bends upon excessive current, disengages contact, and permits spring to break circuit.

TIME DELAY is usually introduced by an oil dash pot. Trigger normally holds contacts closed; coil is energized by current through breaker. Disc at end of core adheres to bottom of oil cup. If current becomes excessive, pull of coil on core gradually weakens cohesion between disc and oil cup bottom. After preset time, disc is released. Core is pulled up, hits trigger, allows spring to open contacts.

Contacts

Spring

Trigger

Coil

Core

Scale

Adjustment

Oil

Disc

SMALL BREAKER UNITS, magnetic and thermal, are made for mounting into complete circuit breaker panelboards or load centers.

Operating handle

Breaker (Direct Mounting)

Breaker (Remote Mounting)

OIL CIRCUIT BREAKERS up to 2500 volts of small current-carrying capacity may be mounted directly to switchboard. Larger units are remotely mounted up to 50 ft from operating handle, connected mechanically by rods and cranks. Contacts are made and broken under oil for thorough quenching of arc.

Insulator

Stationary Contact

Moving Contact

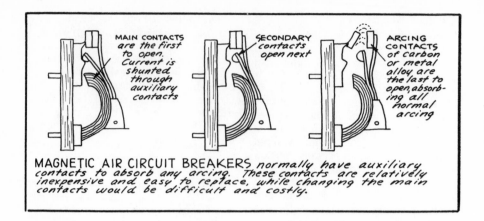

MAIN CONTACTS are the first to open. Current is shunted through auxiliary contacts

SECONDARY contacts open next

ARCING CONTACTS of carbon or metal alloy are the last to open, absorbing all normal arcing

MAGNETIC AIR CIRCUIT BREAKERS normally have auxiliary contacts to absorb any arcing. These contacts are relatively inexpensive and easy to replace, while changing the main contacts would be difficult and costly.

ment for homes and in distribution panelboards as control and protection against overload for many types of circuits. Many large breakers are used for controlling large power loads. CB's used for branch circuit protection are constructed to prevent tampering with the current trip point or changing of the time required for operation. Such units are of the air-break type or oil-immersed type. Air circuit breakers may be of open construction for use in switchboards, or of the enclosed unit construction for use in panelboards or individual mounting.

• **Thermal type** air circuit breakers are used primarily for protection against overcurrent. Complete protection against overcurrent, under-current. reverse current, reverse phase and low voltage is generally provided by magnetic type breakers. A typical thermal ACB (air circuit breaker) operates on the principle of expansion of metal when heated. When current in excess of the unit's rating flows through the metal, it expands and opens the circuit. In some CB's the heat is provided by a separate heater element and transferred to the metal. Usually, such units are the large frame sizes and incorporate a magnetic trip device for short circuit conditions.

Thermal ACB's provide overload protection at voltages up to 600 volts ac and 250 volts dc, from 10 to 600 amperes. These units are generally plastic encased, are trip free of the handle and can be mounted in any position.

Thermal breakers for panelboard mounting are made in one-, two- and three-pole units up to 50 ampere size. Multi-pole breakers have a single common handle, and overload on any pole opens all of the contacts. Thermal breakers are also made for mounting in switchboards. These are 100, 225 and 600-amp sizes of breakers, in two- and three-pole units, for use on feeders or branch circuits. Breakers of these sizes are also commonly used in individual enclosures.

• **Magnetic air circuit breakers** provide opening of their contacts, when current exceeds the rated value, by means of a magnetic coil or solenoid. Generally, a magnetic ACB has three sets of contacts: one set for the main current flowing through the breaker; secondary contacts which open after the main contacts are opened; and the arc contacts which are last to open. Contacts in these breakers receive heavy duty operation and are either solid silver or silver composition.

Magnetic breakers are used on heavy mains, on large feeders, as tie breakers and other heavy duty applications. Typical magnetic breakers

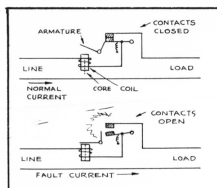

FULLY MAGNETIC BREAKER

has no thermal element and is thus unaffected by ambient temperature. Greater-than-rated currents magnetize coil, draw iron core into coil, and attract armature, opening breaker contacts. Short circuits cause instantaneous opening of contacts; overloads draw core into coil at speed proportional to current, time delay being introduced by fluid or other restraining action.

MAGNETIC BREAKER

designed for instantaneous tripping, with no thermal element or provision for time delay, are intended for use with additional overload protection, such as in conjunction with a motor controller as shown.

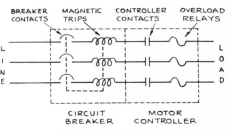

are made for voltages up to 600 volts ac and 750 volts dc, in current ranges up to 6000 amperes ac and 10,000 amperes dc. Small magnetic breakers are widely used up to 460 volts, 50 amps ac and up to 250 volts dc, and are made for panelboard mounting and branch circuit protection like the thermal breakers described.

• **Oil circuit breakers** are another type of circuit breaker, made for both indoor and outdoor applications. In an oil CB, the contacts are submerged in oil to suppress the arcing action when opening. Outdoor OCB's are used for protecting high-voltage transmission and distribution lines, and are mounted in weatherproof enclosures. Indoor OCB's are used on high voltage equipment and distribution lines.

Typical ratings on oil circuit breakers are: 200 to several thousand amperes; voltage ratings up to 290 kv, typical units at 5000 volts, 7500 volts and 15,000 volts. Two-, three- and four-pole units are available with rat-ings from 15,000 to 250,000 kva interrupting capacity. Oil circuit breakers may be manually operated or controlled by electrical solenoid operation.

FOR THE WIDE RANGE of over-current and control applications in circuits rated up to 600 volts, there are basically three types of circuit breakers used:

1. Molded-case breaker is a circuit breaker mechanism assembled as a complete switching and protective device in a supporting and enclosing housing of insulating material. Such units are made in sizes from 10 amps up to 1600 amps, with one, two or three poles.

Molded-case breakers are divided into size groups on the basis of frame sizes. A given frame size embraces a range of ratings of CBs which have the same size molded case, the same rating of contacts and the same current interrupting rating. The continuous current rating of any CB is determined by the rating of its trip element, that value of current at which

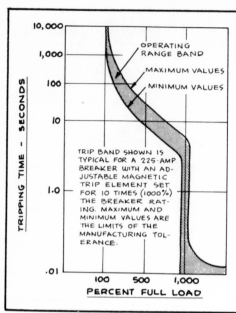

CIRCUIT BREAKER TRIP CURVES describe the tripping action of specific breakers under various loading conditions.

- At or below 100% rated load, CB will not trip. (Moving up from 100% on bottom scale misses operating range band.)

- Between 100% and 1000% full load, time delay is introduced. For example, at 500% full load, CB will not trip unless load persists longer than 9 seconds. (Moving up from 500% intersects operating range band at about 9 seconds.

- At or above 1000% full load, tripping is virtually instantaneous.

Figure labels (left):

TRIPPING TIME – SECONDS

10,000 / 1,000 / 100 / 10 / 1.0 / .01

OPERATING RANGE BAND
MAXIMUM VALUES
MINIMUM VALUES

TRIP BAND SHOWN IS TYPICAL FOR A 225-AMP BREAKER WITH AN ADJUSTABLE MAGNETIC TRIP ELEMENT SET FOR 10 TIMES (1000%) THE BREAKER RATING. MAXIMUM AND MINIMUM VALUES ARE THE LIMITS OF THE MANUFACTURING TOLERANCE.

100   500   1,000
PERCENT FULL LOAD

---

the CB will automatically trip open. The cases of a given frame size are, therefore, physically interchangeable, with the exception of type NI (non-interchangeable) breakers, which have provisions to discourage unauthorized oversizing of circuit protection. The frame size is designated by reference to the largest ampere rating available in that group, e.g., 100-amp frame, 225-amp frame, 400-amp frame, etc. The various manufacturers also use letter designations to establish frame sizes with given current, voltage and interrupting ratings.

Short-circuit interrupting capacities of molded-case circuit breakers vary with frame size and with voltage rating. Molded case CBs may have ICs up to 75,000 amps by themselves and over 200,000 amps when current-limiting fuses are used in conjunction with the CB unit. Molded case CBs are made with either thermal-magnetic or fully magnetic operation, as shown in sketches. Some molded-case CBs—in the range up to 100 amps —have non-interchangeable trip mechanisms of fixed operating characteristics. Others—in the larger sizes—

are available with adjustable magnetic trip elements for selecting the value of short-circuit current at which the CB will trip. Adjustable trip elements may be sealed in the CB or may be part of interchangeable assemblies which can be used in CBs designed for such application.

For application in motor branch circuits, molded-case breakers are available without the time-delay trip element. Such CBs have only the instantaneous trip element to provide short-circuit protection when used in conjunction with overload relays in motor starters. In such application, a 50-amp CB would not trip until the current got up around 500 amps (10 x rating) to operate the short-circuit trip element. Any circuit overcurrent due to motor overload would actuate the runing overload relays in the motor starter.

2. Power circuit breakers, also called "large air breakers," are heavy-duty switching and protective devices with high interrupting capacities. These breakers have rugged contact assemblies, sturdy mechanical construction and either manual or electrically powered operation. They are available either as individually mounted units—in wall or floor mounting enclosures—or in switchboards or load-center substations.

Power CBs can be mounted either stationary or with drawout construction. In the stationary type, the CB is bolted to the switchboard bus and secured to the enclosing structure. In the drawout type, the CB is mounted on a simple racking mechanism which permits easy disconnecting of the CB from the hot buses and withdrawal from the enclosure.

Power CBs are available in ratings from 15 amps up to 4000 amps, continuous current, in ratings of 240, 480 and 600 volts. Interrupting ratings vary with size and voltage from 15,-000 amps up to 150,000 amps asymmetrical (and beyond 200,000 amps in combination with suitable current-limiting fuses). Power CBs provide for circuit breaker application above the range of molded-case breakers (1200 amps). And even within the range of ratings of molded-case CBs, power CBs are suited to applications requiring interrupting capacity higher than that available in molded-case units and to use where high frequency of switching operation demands a heavy duty device.

3. High voltage breakers provide protection and control in circuits operating above 600 volts. There are both oil-immersed units and units with the contacts operating in air. Although the oil type has lower first cost, it does require maintenance of the oil. The air or oilless type has become the more popular of the two for circuit breaker applications up to 15 kv, and even above. Air power breakers provide load switching, short-circuit protection and overload protection, with electrically powered operation for local or remote control of the breaker.

CB Selection

Selection of a CB is based on the following:

1. Current Rating—The continuous current rating of a circuit breaker is, when applied in a circuit, the same as its trip rating. Usually, a breaker is selected to have a trip setting which will protect circuit wires in accordance with their safe allowable current carrying capacity (as given in Tables 310-12 to 310-15 in the National Electrical Code). That is, the CB will open automatically if the current reaches a value which could produce a dangerous temperature in the conductor or its insulation.

In cases where a CB is protecting a motor branch circuit or feeder or a motor-and-lighting feeder, the trip setting may be higher than the current rating of the wire to permit motor starting current to flow without tripping-out the CB. Such oversizing is necessary and recognized by the Code. In motor circuit applications, time-delay overload protective devices usually incorporated in the motor controller, will protect the conductors as well as the motor against overloads up to and including stalled rotor condition of the motor. Then the CB will open the circuit on higher values of fault current, such as grounds or shorts.

Another case of permissible oversizing of CBs for protection is that of protecting transformers. The Code permits use of a CB with trip setting up to 250% of the full-load current of the transformer, to allow short-time overload operation within the design capacity of the transformer.

A CB provides effective control switching and performs easily as a load-break switch for power and lighting circuits. A CB may also be used as a motor controller if it has an ampere rating at least equal to the motor full-load running current. And it may be used as a motor disconnect switch if it has a rating at least equal to 115% of the motor nameplate running current.

SELECTION of a circuit breaker should take into account the two functions it provides in the way of overcurrent protection: **1.** Protection against operating overloads—too many lamps, or appliances, or motors connected to the circuit, so that the current drawn by the total connected load is in excess of the safe value of current which the conductors can carry; and **2.** Protection against fault currents—insulation failures to ground and short circuits.

A breaker can safely and automatically open values of fault current up to the interrupting rating of the contacts. The interrupting rating of a breaker is the maximum value of fault current which the breaker is capable of opening without damaging or destroying itself in the process. Both of the above two functions of a breaker should be correlated to the load on a circuit and the maximum short-circuit current available.

A circuit breaker with time delay operation in its overload range (up to about 10 times the continuous current rating of the unit) can be used to provide both running overload protection and short-circuit protection for motor branch circuits. In such an application, the rating of the CB must not be more than 125% of the motor nameplate current for motors rated for a 40 deg C rise and not more than 115% of motor nameplate current for other motors.

A single circuit breaker can provide all of the functions normally provided by a combination motor controller. The CB can be used to start and stop the motor, will open the circuit if the motor becomes overloaded or if a ground fault or short circuit develops in the circuit. The CB can also provide disconnection required by the Code.

Interrupting Rating—In selection of any circuit breaker, extreme care should be taken to make sure that the unit has sufficient short-circuit interrupting ability to handle the amount of fault current which would flow if a dead short circuit developed on the load terminals of the breaker. In any circuit, such a short-circuit on the

CB load terminals will produce the highest current the CB might be called upon to interrupt. In any particular case, the amount of current depends upon the capacity of the system which supplies the circuit, i.e., in general, the kva rating of the transformer supplying the circuit and the total impedance in the circuit from the transformer to the short.

For instance, a CB with a maximum interrupting rating of 10,000 amps must not be used at a point in an electrical system where, say, 12,000 amps could flow if a short circuit developed. And, it must be noted, this attention to adequate interrupting capacity is required even though the particular CB has a suitable continuous current rating (say, 100 amps) for the load fed through it.

Voltage Rating—Selection of any CB must assure that the breaker has a voltage rating suitable for the voltage level at which the circuit operates.

Number of Poles—The number of poles required in a circuit breaker can be determined from the circuit and the load conditions. As a protective device, any circuit breaker must have a current-responsive trip unit in each ungrounded conductor of the circuit. The NE Code further states that an overcurrent device must not be placed in any permanently grounded conductor, except in the following cases:

1. Where the device simultaneously opens all conductors of the circuit—the ungrounded conductors at the same time as the grounded conductor. Such would be the use of a 3-pole CB in a 120/240-volt, 3-wire, single-phase circuit—with one pole of the CB in each ungrounded conductor and one pole in the grounded neutral wire. But such application is uneconomical (a 2-pole CB with one pole for each hot wire is adequate and actually better).

2. Where the device is used for running overload protection in the grounded leg of a 3-phase delta sup-

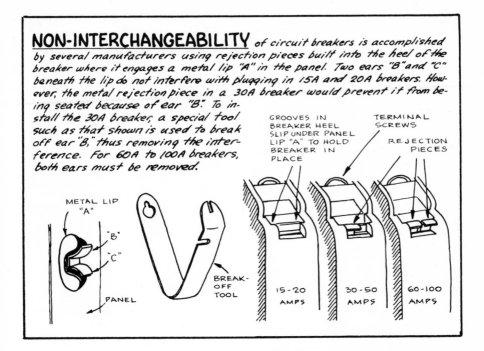

NON-INTERCHANGEABILITY of circuit breakers is accomplished by several manufacturers using rejection pieces built into the heel of the breaker where it engages a metal lip "A" in the panel. Two ears "B" and "C" beneath the lip do not interfere with plugging in 15A and 20A breakers. However, the metal rejection piece in a 30A breaker would prevent it from being seated because of ear "B". To install the 30A breaker, a special tool such as that shown is used to break off ear "B," thus removing the interference. For 60A to 100A breakers, both ears must be removed.

GROOVES IN BREAKER HEEL SLIP UNDER PANEL LIP "A" TO HOLD BREAKER IN PLACE

TERMINAL SCREWS

REJECTION PIECES

METAL LIP "A"

"B"

"C"

PANEL

BREAK-OFF TOOL

15-20 AMPS

30-50 AMPS

60-100 AMPS

ply when three overload devices are required by the Code (Sec. 430-37, Note to Table). Such would be the use of a 3-pole CB in a 3-wire, 3-phase circuit to a motor, tapped from a corner-grounded delta system.

CB selection often involves a choice between use of a number of single-pole devices and use of one multi-pole device. The basic rule requires a 3-pole CB (and not three single-pole CBs) for circuits to 3-phase motors. For 240-volt circuits from a 120/240-volt, 3-wire, single-phase system, either a 2-pole CB or two single-pole CBs may be used. Single-pole CBs may also be used (and are recommended) for protecting the un-grounded conductors of 3-phase, 4-wire circuits serving lighting and/or appliance loads which are connected from a hot leg to neutral.

Service Equipment—Service disconnect and overcurrent protection may be provided by circuit breakers. A manually operated CB may be used as a service disconnect if it is approved for use as service equipment

and for the conditions of application. In addition to being manually operable, a service circuit breaker may also be equipped for electrical operation, such as to provide remote control of the CB.

Where a single main disconnect is required for service, a single multi-pole CB may be used. If the six subdivision allowance of the Code (Sec. 230-70 g) is permitted, the disconnect may consist of up to six CBs. This may be six single-pole CBs or six multi-pole CBs. And two or three single-pole breakers, connected on different ungrounded phase conductors of multi-wire circuits, may be considered to be a single multi-pole breaker if they are equipped with handle ties or have their individual handles within 1/16-in proximity.

A circuit breaker must be capable of being closed and opened by hand, without employing any other source of power, although normal operation may be by other power such as electrical, pneumatic and the like.

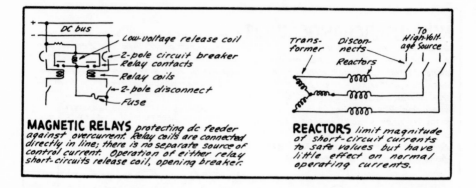

MAGNETIC RELAYS *protecting dc feeder against overcurrent. Relay coils are connected directly in line; there is no separate source of control current. Operation of either relay short-circuits release coil, opening breaker.*

REACTORS *limit magnitude of short-circuit currents to safe values but have little effect on normal operating currents.*

Non - Interchangeability — Circuit breakers used in lighting and appliance branch circuits in residential and other occupancies where the conditions of maintenance and supervision do not assure that CBs will be maintained at proper ratings for the size of wires protected must be Type NI (Non-Interchangeable) breakers. These breakers and the bussing and mounting provisions in panelboards must prevent (or make difficult) the replacing of a CB with one of higher ampere rating. Non-interchangability is provided among three classifications: 0 to 20 amps, 21 to 50 amps and 51 to 100 amps. For instance, a 15- or 20-amp CB cannot readily be replaced by a 30-amp or higher rated CB. Or a 30-amp CB cannot be replaced easily by a 70-amp CB.

IN ADDITION to fuses and circuit breakers, there are other devices which serve protective purposes in electrical circuits and equipment. These include: thermal overload release devices, thermal overload relays, magnetic overload release devices, magnetic overload relays, thermal cutouts and lightning arresters.

Thermal Devices

• **A thermal overload release device** is a mechanism used in many manually-operated switches and circuit breakers to act upon the holding catch of the contact when the current gets too high. The thermal element physically trips open the contacts when it expands due to heating.

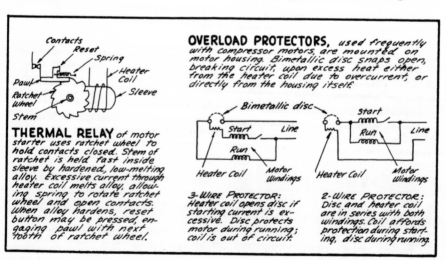

THERMAL RELAY *of motor starter uses ratchet wheel to hold contacts closed. Stem of ratchet is held fast inside sleeve by hardened, low-melting alloy. Excessive current through heater coil melts alloy, allowing spring to rotate ratchet wheel and open contacts. When alloy hardens, reset button may be pressed, engaging pawl with next tooth of ratchet wheel.*

OVERLOAD PROTECTORS, *used frequently with compressor motors, are mounted on motor housing. Bimetallic disc snaps open, breaking circuit, upon excess heat either from the heater coil due to overcurrent, or directly from the housing itself.*

3-WIRE PROTECTOR: Heater coil opens disc if starting current is excessive. Disc protects motor during running; coil is out of circuit.

2-WIRE PROTECTOR: Disc and heater coil are in series with both windings. Coil affords protection during starting, disc during running.

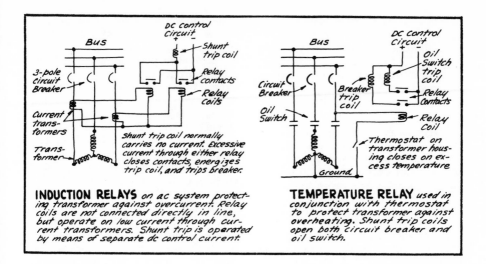

INDUCTION RELAYS *on ac system protecting transformer against overcurrent. Relay coils are not connected directly in line, but operate on low current through current transformers. Shunt trip is operated by means of separate dc control current.*

TEMPERATURE RELAY *used in conjunction with thermostat to protect transformer against overheating. Shunt trip coils open both circuit breaker and oil switch.*

• **Thermal relays** are commonly used in magnetic switches used for motors. They provide overload protection for the motor. Unlike thermal release devies, they are not generally used in circuit breakers.

Overload protection by a thermal relay has inverse-time characteristics, i.e., the higher the value of overload current, the shorter the time in which the relay will open the circuit. A thermal relay does not itself provide interruption of the main full-load current. It breaks contact in an auxiliary circuit, de-energizing a coil which holds the switch contacts in the closed position. The switch contacts then open, interrupting the main current. Thermal relays are made in both self-resetting and manual-resetting types.

• **The thermal cutout** is another overload protective device, with inverse-time characteristics.

Particular applications of thermal cutouts are: protection of motors under excessive overloads, protection against single-phasing of polyphase motors (the condition when one of the three leads to a three-phase motor opens, and the remaining two leads

form a heavily overloaded single-phase operating condition which keeps the motor running), and other cases in which particular protection is needed against excessive overloads in motors.

A typical thermal cutout contains a heating coil, a fusible link and a spring contact arm. Line current flows through the coil. If this develops excessive heat due to excessive current flow, the fusible link will melt and the contact arm will spring back to open the line. In this action, inverse-time characteristics obtain: if the coil heats quickly due to a heavy overload, the line will be opened fast; if a light overload develops, the heating action and, consequently, the opening of the device will take longer.

Magnetic Devices

An electrically energized solenoid coil which controls the movement of an iron plunger or armature is the essential element in a magnetic relay or magnetic overload release device. The coil may be directly connected in series in the line to be protected or it may be fed from a current transformer with its primary connected in series with the line. A magnetic release device may be set to provide opening of the line at a particular value of overload current by adjusting the travel of the plunger which acts upon the contact mechanism.

Magnetic relay and magnetic release devices may incorporate mechanical elements which provide a particular timing of operation. Instantaneous devices function as soon as any overload exists. Other units have inverse-time or definite-time characteristics.

A magnetic relay acts upon an auxiliary circuit which controls the contact mechanism of the switch. A release device, on the other hand, acts directly upon the contact mechanism. Magnetic release devices are used in circuit breakers; magnetic relays are used in switches and breakers.

Lightning Arresters

Protection against high voltage lightning surges in outdoor electric lines is provided by lightning arresters. During lightning storms, very high voltages may be built up in outdoor lines and carried along the lines to electrical equipment, either indoors or outdoors. Lightning arresters are used to carry off the dangerous voltages before they can reach and damage equipment, or cause fire or other damage.

A lightning arrester must provide a path for the lightning to get to ground, quickly and without flashing. It must prevent flashing of the lightning over the equipment insulation, and it must prevent power current from following the path taken by the lightning.

Many forms of lightning arresters are made for both indoor and outdoor

applications. A typical arrester for use on lines up to 15 kv rating consists of a spark gap in series with a valve element. The gap insulates the line from ground. When lightning strikes, the surge jumps the gap within the housing of the arrester and travels quickly through the low resistance valve element which easily passes the surge but offers high resistance to the power current which tries to follow the voltage surge. Such an arrester is a valve type arrester.

Another type of lightning arrester used on lines from 3 to 18 kv rating is the expulsion type. This consists of a series of short spark gaps between line and ground. An isolating gap in open air is located at the top of an expulsion arrester, and a second gap element is located in the arrester housing. A lightning surge jumps the isolating gap first and then surges over the series gaps in the arrester housing, running off to ground.

Secondary lightning arresters are used on distribution circuits up to 600 volts. These are often similar to valve type arresters used on high voltage lines. They carry off dangerous lightning strikes, using gap and valve element action.

In many large distribution systems, surge protective equipment is used to limit the value of surge and short circuit currents. Basically, surge protection is provided by coils of wire which act to oppose current flow through themselves. They are also used with lightning arresters to shunt current from the path to equipment to the path through the arrester. The current tends to take the lower resistance path through the arrester when a surge takes place. When surge protective coils are used to limit short circuit current, they are called reactors.

Insulating Materials

ALTHOUGH current-conducting materials—such as copper or aluminum wires, busbars, connectors, etc.—are the primary elements of electrical equipment, circuits and systems, the practical application of electricity would be impossible without the use of insulating materials. These are the non-current-conducting materials—or dielectrics—which are used to enclose current carrying parts, assuring proper operation of circuits and minimizing hazards of electricity.

The following is a breakdown of the common types of insulating materials used in the construction and maintenance of electrical systems and the repair of equipment.

Tapes

A wide variety of tapes is used for insulating. Some are used in electrical construction; some, in equipment insulation. Some tapes are adhesives; others are not.

Typical tapes used for splicing and terminating wires and cables include friction tape, rubber tape and plastic tape. Friction tape consists of fabric to which a strong dielectric is bonded by impregnation. And the rubbery impregnation provides a firm adhesive hold. Available in a number of quality grades, friction tapes are used for general-purpose and high-voltage splices and wraps and for wrapping over splicing compounds.

Rubber tapes or splicing compounds, as they are commonly called, consist of rubber compounds made into thin tape form. They have high insulating quality and tensile strength. They are self-conforming to irregular surfaces, have adhesive tackiness and readily mold into a solid insulating mass. It is common practice to use friction tape over rubber tape for added mechanical protection.

Plastic tapes are made of special plastics with high dielectric strength, great mechanical strength and elongation characteristics. These tapes are adhesives and provide ready molding over irregular shapes. Plastic tapes have come to be used in place of rubber-and-friction tape splicing. They offer speed and ease for splicing in tight quarters. Typical plastic tape is highly resistant to flame, abrasion, sunlight, moisture, alkalies, solvents, oil and many acids. Plastic tapes are made in a number of thicknesses for various abrasion resistances.

Another type of splicing tape is varnished tape. This tape provides electrical varnish protection for high puncture resistance. A conformable varnish tape combines varnish insulation with a plastic tape, for use on motor connections or for splicing varnished cambric cables. Varnished cambric tape with adhesive backing is used for the same application.

Still other insulating tapes include special high-voltage tapes for use on rubber, neoprene or plastic insulated cable splices or connections, and glass cloth tapes for use on asbestos or glass insulated wires in high temperature applications.

Cable fireproofing tape is a dry tape made of asbestos. It may be silicone-impregnated to repel water or may be treated with waterglass for moisture resistance. Such tape is commonly used to bind individual conductors of feeders into cabled assemblies in pull boxes and switchboard rooms. The tape wrapping protects each set of feeder conductors and provides ready identification of conductors.

Electrical tapes are used in many ways in the rewinding and repair of motors and transformers. Tapes are used for fastening the turns of coils, holding windings in place, reinforcing slot liners and a wealth of other major and minor taping details. Such tapes include: crepe paper tapes, glass cloth tapes, cotton cloth tapes and plastic film tapes.

Varnishes and Paints

Insulating by means of varnishes and paints is used in some phases of electrical construction work and in motor and transformer repair operations. Insulating varnishes are solutions of resins and drying oils in suitable solvents. They are used to coat, to bond, to fill, to seal or to protect electrical equipment against heat, oil, chemicals or mechanical forces.

Basically, varnishes can be divided into several different types. Baking varnishes are dried by heat in baking ovens. Such varnishes are generally tougher and more effective insulations and are used on equipment for heavy duty applications. Air-drying varnishes are used where baking is not

Class	Hottest spot temperature (deg C)	Permissible rise (deg C)	Materials Used
O	90	35 - 45	Cotton, silk, paper and similar organic materials not immersed or impregnated with dielectric.
A	105	50 - 60	Same as Class O, but immersed or impregnated with dielectric; moulded and laminated materials with cellulose filler or resins; films and sheets of cellulose derivatives; varnishes (enamel) applied to conductors.
B	130	70 - 80	Mica, asbestos fiber glass and similar inorganic materials in built-up forms, with organic binding substances.
C	Not yet established		Mica, asbestos fiber glass and similar inorganic materials in built-up forms.
H	180	100 - 120	Same as Class C in built-up forms with silicone compounds or equivalent as binding substances.

INDUSTRY STANDARDS define classes of insulation for electrical machinery in accordance with permissible operating temperatures. Permissible rise refers to rise in temperature above ambient air until the temperature is steady with the machine running fully loaded. The range shown is due to differences in measuring methods.

ENCAPSULATING *of coils involves first wrapping wire coil with glass tape or cloth (above), then impregnating and encasing coil in a mould with an insulating compound, plus a finish shell of resin (below).*

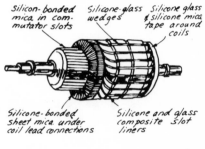

CLASS H INSULATION, *which permits higher operating temperatures than all other standard classes used for general industry applications, uses silicones and other inorganic materials as illustrated by the motor armature below.*

Silicon-bonded mica in commutator slots

Silicone-glass wedges

Silicone glass & silicone mica tape around coils

Silicone-bonded sheet mica under coil lead connections

Silicone and glass composite slot liners

A Class H-insulated stator uses, in addition, silicone oil and grease for bearings, silicone sleeving over silicone-rubber-insulated magnet wire, and silicone-treated tying cord.

possible or practical. They are not as effective or long lasting as the baking type. Finishing varnishes are highly resistant to oil and gasoline, making them well suited to use as a final coating over other insulations. They dry to a hard, smooth finish. Varnishes are made in clear and black types, each of which has particular application advantages.

The most common method of applying varnish is by dipping the parts to be insulated in a tank containing the varnish. Vacuum or pressure impregnation may also be combined with the dipping process to assure full penetration of the varnish into the winding or part being insulated.

Insulating enamels are general purpose insulating finishes for coils, end windings and armatures, where resistance to chemicals, oil, acids or moisture is very important.

Insulating Compounds

A number of compounds are available for use as insulation. One type is known as filling compound and is used for filling large openings in coils, ballasts and potheads—where thin varnish would not be suitable. Typical compounds for impregnating or filling are poured, after heating, into the work.

Another insulating compound comes in the form of a roll of tape, called insulating putty. Strips of this putty are used to mold an insulating jacket around splicing connectors of the bolted type. The putty is readily formed to any shape and then an outer wrapping of electrical tape is applied to seal the jacket in place.

Still other compounds are available in liquid form for making cast insulating jackets on splices in the field. Such liquid resin compound is available in premeasured quantities for specific sizes of splices. When mixed, the resin sets in about 30 minutes to a solid, firm insulating mass. It can be used for insulating or sealing. This compound is also sold in special splicing kits which contain all the materials to make cast splices, including the resin and the molds.

Transformers

A transformer is a device by means of which electrical energy can be transferred from one alternating-current circuit to another. The basic type of transformer has no moving parts. Usually, use of a transformer involves a change in voltage when energy is transferred. If the output voltage of a transformer is higher than the input voltage, the device is called a "step-up" transformer. If the output voltage is lower than the input voltage, the unit is called a "step-down" transformer. If the output and input voltages are the same, the transformer is called a "one-to-one" or "isolating" transformer.

Transformers cannot operate on direct-current; direct-current would burn out the wire in the transformer. They derive their operation from the fundamental characteristics of alternating-current.

A transformer is not a piece of electrical utilization equipment. It does not convert electrical power to mechanical power like a motor or change electrical power into light like a lamp. It is, instead, a piece of electrical distribution equipment, making possible convenient, extensive and effective application of alternating-current. As a static device, it requires little maintenance and is efficient and economical.

Operation

Operation of a transformer is based on the fundamental characteristics of ac electrical energy. A series of causes and effects are involved, as follows:

1. When alternating-current flows through an electrical conductor, a magnetic field continuously builds-up and collapses around the conductor, in time with the frequency of the ac.

2. If the conductor is wound into a coil of many turns of wire, the magnetic fields around the individual turns are added together to form a single, concentrated field around the coil.

3. If this coil is placed around a piece of iron, the magnetic field is greatly strengthened and builds-up and collapses around the entire piece of iron.

4. If a second coil is placed around the piece of iron, the pulsating magnetic field produced by the first coil will cut across the second coil, as it builds and collapses due to current alternations in the first coil.

5. Interaction between the pulsating magnetic field and the second coil will set up a voltage in the second coil—a

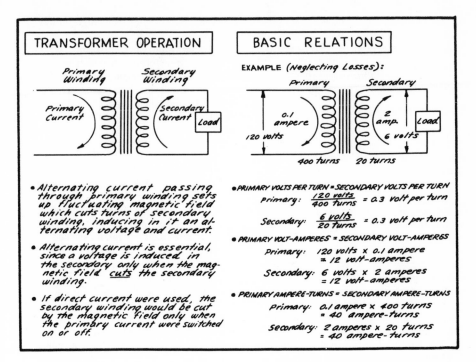

TRANSFORMER OPERATION

Primary Winding   Secondary Winding

Primary Current   Secondary Current   Load

- Alternating current passing through primary winding sets up fluctuating magnetic field which cuts turns of secondary winding, inducing in it an alternating voltage and current.

- Alternating current is essential, since a voltage is induced in the secondary only when the magnetic field _cuts_ the secondary winding.

- If direct current were used, the secondary winding would be cut by the magnetic field only when the primary current were switched on or off.

BASIC RELATIONS

EXAMPLE (Neglecting Losses):

Primary   Secondary

0.1 ampere   120 volts   2 amp.   6 volts   Load

400 turns   20 turns

- PRIMARY VOLTS PER TURN = SECONDARY VOLTS PER TURN

  Primary: $\dfrac{120 \text{ volts}}{400 \text{ turns}} = 0.3$ volt per turn

  Secondary: $\dfrac{6 \text{ volts}}{20 \text{ turns}} = 0.3$ volt per turn

- PRIMARY VOLT-AMPERES = SECONDARY VOLT-AMPERES

  Primary: 120 volts × 0.1 ampere = 12 volt-amperes

  Secondary: 6 volts × 2 amperes = 12 volt-amperes

- PRIMARY AMPERE-TURNS = SECONDARY AMPERE-TURNS

  Primary: 0.1 ampere × 400 turns = 40 ampere-turns

  Secondary: 2 amperes × 20 turns = 40 ampere-turns

---

phenomenon similar to previously described "motor action."

Thus, electrical energy applied to the first coil or transformer primary is transferred to the second coil or transformer secondary. There is no current flow from the first to the second coils. Instead, energy is transferred from the primary to the secondary by induction—that electrical effect by which voltage is set up in conductors when there is relative motion between the conductors and a magnetic field. In a generator, voltage is set up by the same principle. In the transformer, the rise and fall of the magnetic field set up by the primary produces relative motion between the field and the conductors of the secondary. Voltage is induced in the secondary as a result.

In a transformer, if the number of turns in the first or primary coil is the same as the number of turns in the second or secondary coil, the voltage produced in the secondary would be equal to the voltage which was applied to the primary. If the number of turns on the primary is greater than the number of turns on the secondary, the secondary voltage will be lower than that applied to the primary. If the number of turns on the primary is less than the number of turns on the secondary, the secondary voltage will be greater than the primary voltage. The ratio between number of primary and secondary turns, therefore, determines whether the transformer is a "step-up" or "step-down" transformer and also determines the amount of voltage change between the two windings.

The transformer winding to which energy is supplied is called the "primary winding." The winding from which energy is taken to feed an external circuit is called the "secondary winding."

In the operation of any transformer, the amount of power supplied from the secondary is never greater than the amount of power applied to the primary. In fact, there are some power losses in the transformer so that the power output is always less than the power input by the amount of the losses. This is true whether the input voltage is stepped-up, stepped-down or maintained the same on the secondary as in a one-to-one transformer. The lower the losses of any transformer, the higher the efficiency.

Construction

Transformers vary widely in their construction characteristics depending upon size and type. Depending upon the arrangement of the iron core and the windings, a transformer may be either a core type or a shell type.

In core type transformers, the core is made of sheet-steel laminations arranged in the form of a hollow square. The two windings are wound around two opposite legs of the core, with the high voltage winding wrapped over the low voltage winding and insulated from it.

In shell type transformers, the core is also made of sheet-steel laminations but is in the form of a squared-off figure 8. The windings are made like pancakes and arranged around the inside leg of the core. The winding arrangement consists of coils of the low voltage winding

adjacent to coils of the high voltage winding.

Another type of transformer is the Spirakore transformer in which the core is made of steel strip or ribbon which is wound around the preformed transformer windings. This type of core has a number of advantages including minimum losses and a tighter construction with improved magnetic characteristics.

Hipersil cores are made of special steel by winding the strip on a form and impregnating it with molten glass. The core is then cut in two to assemble it with the windings. Like the Spirakore, it is a better core and can be made faster than a core made of hand-stacked laminations.

Two other cores are the round-wound and the bent-iron types. In the first, strip steel of varying widths is wound on a form to produce a core cross-section that is almost a circle around which the windings can be tightly wound. In the bent-iron type, instead of a wound core of strip steel, steel strips or laminations are bent around the windings to produce a sort of shell-type core.

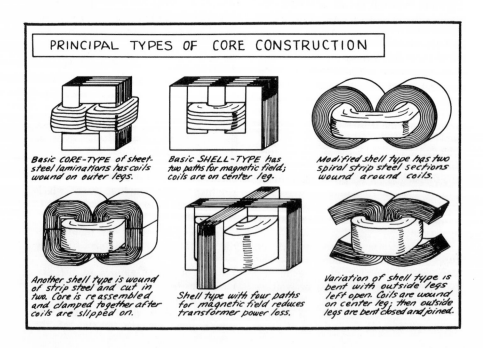

PRINCIPAL TYPES OF CORE CONSTRUCTION

Basic CORE-TYPE of sheet-steel laminations has coils wound on outer legs.

Basic SHELL-TYPE has two paths for magnetic field; coils are on center leg.

Modified shell type has two spiral strip steel sections wound around coils.

Another shell type is wound of strip steel and cut in two. Core is reassembled and clamped together after coils are slipped on.

Shell type with four paths for magnetic field reduces transformer power loss.

Variation of shell type is bent with outside legs left open. Coils are wound on center leg; then outside legs are bent closed and joined.

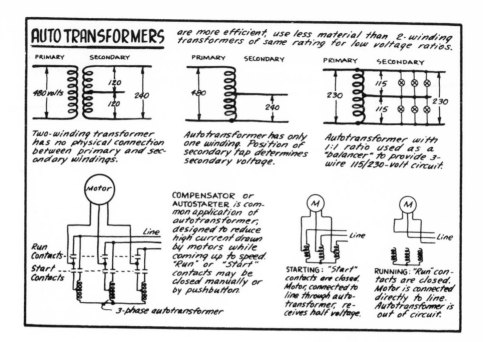

AUTO TRANSFORMERS

are more efficient, use less material than 2-winding transformers of same rating for low voltage ratios.

PRIMARY   SECONDARY

480 volts — 120 — 240 — 120

Two-winding transformer has no physical connection between primary and secondary windings.

PRIMARY   SECONDARY

480 — 240

Autotransformer has only one winding. Position of secondary tap determines secondary voltage.

PRIMARY   SECONDARY

230 — 115 — 230 — 115

Autotransformer with 1:1 ratio used as a "balancer" to provide 3-wire 115/230-volt circuit.

Motor

Line

Run Contacts

Start Contacts

3-phase autotransformer

COMPENSATOR or AUTOSTARTER is common application of autotransformer, designed to reduce high current drawn by motors while coming up to speed. "Run" or "Start" contacts may be closed manually or by pushbutton.

M

Line

STARTING: "Start" contacts are closed. Motor, connected to line through autotransformer, receives half voltage.

M

Line

RUNNING: "Run" contacts are closed. Motor is connected directly to line. Autotransformer is out of circuit.

A TRANSFORMER in operation develops heat due to energy lost during voltage transformation. Although this heat is not great in the case of small transformers, it becomes appreciable and requires special provisions for its dissipation in larger transformers. Because the removal of heat is so important to proper operation of transformers and greatly influences size and construction, transformers are classified according to the method used for cooling and insulating:

Dry type transformers are cooled by circulation of air over the outside of the transformer housing. These are generally smaller units, such as the common wall-mounting lighting transformer or instrument transformers. They may be used indoors without a fireproof vault, a requirement in the case of oil-immersed transformers.

Oil-immersed, self-cooled transformers are cooled by natural circulation of oil through the core and windings. The oil carries the heat to the surface of the housing where it is carried away by the air. Units with power ratings of greater than 40 kvs do not have enough heat dissipating surface on their housings. Such units require fluted sides to their housings or external tubes through which heated oil flows from the top to the bottom of the housing, giving up the heat to naturally circulated air which is constantly in contact with the tubes.

Oil - immersed, forced - air - cooled transformers are similar to the oil-immersed, self-cooled type in that the core and windings are immersed in oil and the oil carries heat from the windings to cooling surfaces such as the housing and tubes or radiators on the housing. However, they have an additional cooling aid —forced circulation of the air which contacts the cooling surfaces and carries the heat away. This forced circulation of the air is usually accomplished by means of fans which are mounted on the outside of the housing and which direct their air streams against the tubes or radiators.

Oil-immersed, water-cooled transformers have their cores and coils immersed in oil within the housing and are cooled by water which is circulated through pipe coils immersed in the oil with the transformer windings. Precaution must be taken against freezing of this water in cold weather.

Oil - immersed, forced - oil - cooled transformers have their cores and coils immersed in oil and are cooled by forced-oil circulation through external radiators. A pump is used in these units to force the oil up through the windings and back to the bottom of the housing through the external cooling tubes. Fans may also be used with these units to provide forced-circulation of the air which removes heat from the oil in the tubes.

Air-blast transformers are essentially dry-type units in which air circulation is forced through the cores and coils. These units, although limited to use at not over 25,000 volts, find considerable application in substations where the use of oil would constitute a fire hazard.

"Breathing" is a term used to describe the action whereby air is drawn into transformers due to expansion of the oil in the transformer. When the transformer is warm, the oil expands and forces gas out of the transformer housing; when the transformer cools, air is drawn in.

Breathing produces several undesirable effects in transformers. If moisture is drawn in with the air, the insulation properties of the oil are adversely affected. Oxygen will oxidize the oil, forming a sludge which may clog the oil ducts and even cause burnout of the transformer. Oxygen also increases the chances that flash-overs within the unit may cause fire or explosion. Because of these possibilities, transformers must either be completely sealed-off with covers, gaskets and sealing compounds or they must be equipped with breathers.

Breathers are devices which are attached to transformers to prevent the harmful effects of moisture and oxygen on the oil. They contain a dehydrating agent which eliminates moisture; they prevent deterioration of the oil; and they

prevent the formation of explosive mixtures within the transformer.

Mineral oil is the oil generally used in transformers for cooling and insulation. However, synthetic, non-inflammable cooling compounds have been developed to replace oil in applications where fire hazard exists.

Application

Single-phase transformers are used where a single-phase voltage at one level is to be changed to a single-phase voltage of another level. Such transformers have one primary winding and one secondary winding. A typical example of this type would be a 480/240-120 volt unit. A single-phase 480-volt supply is connected to the primary; 240 volts is then available between two of the secondary leads and 120 volts between either of these two leads and a third lead. This is a step-down transformation, which is far more common than step-up in modern electrical systems.

For transforming three-phase voltage at one level to three-phase voltage at another level, either three single-phase transformers or one three-phase transformer may be used. A three-phase transformer weighs less and takes up less space than three single-phase transformers of the same rating. But the use of three single-phase units has a particular advantage: if one phase fails, only that one transformer has to be replaced; a one-phase failure in the case of a three-phase transformer requires replacement of the entire unit of three phases, two of which are not defective.

An autotransformer is a transformer of one winding, with one set of leads connected across the entire winding and another set of leads connected across only part of the winding. Such a unit can be used for either voltage step-up or step-down. Autotransformers require less wire than their two-winding counterparts.

They are invariably small units, with

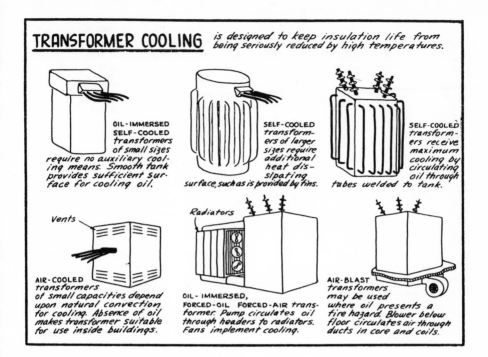

TRANSFORMER COOLING is designed to keep insulation life from being seriously reduced by high temperatures.

OIL-IMMERSED SELF-COOLED transformers of small sizes require no auxiliary cooling means. Smooth tank provides sufficient surface for cooling oil.

SELF-COOLED transformers of larger sizes require additional heat dissipating surface, such as is provided by fins.

SELF-COOLED transformers receive maximum cooling by circulating oil through tubes welded to tank.

Vents

AIR-COOLED transformers of small capacities depend upon natural convection for cooling. Absence of oil makes transformer suitable for use inside buildings.

Radiators

OIL-IMMERSED, FORCED-OIL FORCED-AIR transformer. Pump circulates oil through headers to radiators. Fans implement cooling.

AIR-BLAST transformers may be used where oil presents a fire hazard. Blower below floor circulates air through ducts in core and coils.

moderate ratios of transformation. Common uses for autotransformers are: as "balance coils" to obtain a 3-wire ac system from a 2-wire supply—as an example, to change a 2-wire 240-volt circuit to a 3-wire lighting circuit of 120 volts between each of two hot legs and the neutral which connects to the center of the autotransformer; as starting compensators to start induction motors and synchronous motors.

A disadvantage of the autotransformer derives from the fact that the primary and secondary windings are conductively connected through the one winding to which both primary and secondary leads connect. This constitutes a serious hazard if the voltage is high. Good practice in the use of autotransformers dictates that the low side of the winding be grounded for safety.

ANY bank of 3-phase transformers has its primary and secondary windings connected in one of several

different ways. These ways are familiarly described in industry terminology as "wye-to-wye", "delta-to-delta", "wye to-delta" and "delta-to-wye." These are the most common types of 3-phase transformer connections; and because the descriptive phrases used occur so frequently in conversation on electrical systems, some understanding of the terms should be had.

In the expression "wye-to-wye", the first "wye" indicates that the three windings which are fed by the 3-phase supply line are connected with a particular relationship among the three ac voltages. The second "wye" indicates that the three windings which make up the 3-phase secondary and feed the load are also connected with the same particular relationship among the three voltage phases. In the expression "delta-to-delta", the "delta" terms indicate that both the primary and secondary of the 3-phase transformer bank have the same relationship among the voltage phases, a different relationship, however, than that which exists among the voltage phases in a wye connection. In the expression "wye-to-delta" or in "delta-to-wye", the terms are different to indicate that the primary windings of the transformer bank are connected with one relationship among voltage phases, while the secondary windings are connected with another relationship.

Another not uncommon connection for transforming 3-phase power from one voltage level to another is called the "vee-to-vee" or "open delta" connection. Using this type of connection for both the primary and secondary sides of the transformer bank, it is necessary to use only two windings on each side. Instead of using three single-phase transformers, therefore, only two single-phase transformers will do the job and still provide 3-phase transformation.

A special connection called the "Scott" or "T" connection also makes possible the use of only two single-phase transformers for 3-phase voltage transformation. The connection can also be utilized for changing 3-phase power to 2-phase power and vice versa.

Tap changing on transformers consists of changing the connections to the transformer winding to alter the transformer's ratio of voltage transformation Some means is usually provided on transformers to allow tap changing to increase or decrease the step-up or step-down of the voltage, as voltage-drop might require in many cases. Simply moving a line connection from one terminal to another on the transformer is one manual way of changing taps. Such a procedure, however, is time-consuming and involves disconnecting and disassembling the transformer. In smaller transformers, therefore, an insulated switch which sticks up out of the transformer housing allows quick and convenient changing of taps, although the circuit is opened momentarily.

Tap changing on large power transformers can be done while the unit is operating, without opening the circuit.

This allows regulation of voltage on feeders, automatically and without the use of induction regulators. And in many industrial applications, control and adjustment of voltage is highly desirable.

Most transformers used today, for electrical transmission and distribution, are designated "constant-potential" (or "constant-voltage") transformers. This means that the voltage across the secondary winding remains fairly constant for any change in load, the secondary current being the variable which changes with load. There are, however, certain applications where a constant current is required. A typical case of this is series street lighting. In such applications, it is required that the current feeding the series load be maintained constant for any value of load. In a series street lighting system, the current through all of the lamps connected in series must be held constant whether lamps are added or taken out of the series circuit. To provide constant-current, a special transformer is used.

A "constant-current" transformer is constructed such that the primary and secondary can move with respect to each other. Either the primary is fixed and the secondary can move, or the secondary is fixed and the primary can move. The transformer itself, under operating conditions, produces magnetic forces within itself to move the primary and secondary closer together or further apart, as required to maintain constant current through the load. While the transformer is operating with a particular load, the two windings maintain fixed positions. If, however, there is a change in the number of lamps on the circuit, the windings will move together or apart whichever is required to bring the current in the circuit back to the desired constant value.

HIGHER distribution and utilization voltages used in modern electrical systems for commercial and industrial buildings have greatly stimulated application of transformers. Increased use of 4160-volt and 13,-

THREE - PHASE TRANSFORMERS

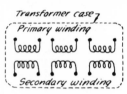

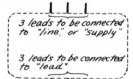

Transformer case

Primary winding

Secondary winding

3 leads to be connected to "line" or "supply"

3 leads to be connected to "load"

A three-phase transformer is the equivalent of three identical single-phase transformers in a single case, wound on a single core. Connections between phases are usually made inside case. Three "line" leads and three "load" leads are brought out of the case.

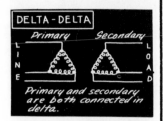

DELTA - DELTA

Primary   Secondary

Primary and secondary are both connected in delta.

DELTA CONNECTION
Line leads

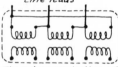

The primary winding of this transformer is connected in "delta". The three winding sections are hooked in tandem, and the 3 line leads are connected as shown.

WYE CONNECTION
Line leads

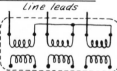

The primary of this transformer is connected in "wye" or "star". One end of each of the 3 windings is hooked together, and the 3 line leads are connected as shown.

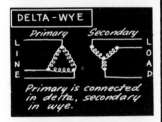

DELTA - WYE

Primary   Secondary

Primary is connected in delta, secondary in wye.

The connections above are usually represented in simplified form as shown below. These simplified diagrams show connections only and do not indicate physical positions of the windings.

DELTA

WYE

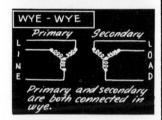

WYE - WYE

Primary   Secondary

Primary and secondary are both connected in wye.

CONSTANT CURRENT TRANSFORMER

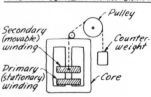

Pulley

Secondary (movable) winding

Counter-weight

Primary (stationary) winding

Core

The "station" type constant current transformer illustrates the basic arrangement of component parts.

SCOTT CONNECTION

3-phase

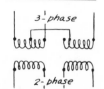

2-phase

The "Scott" or "T" connection is one way to obtain 2-phase power from a 3-phase source or vice versa.

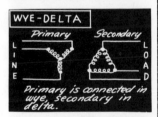

WYE - DELTA

Primary   Secondary

Primary is connected in wye, secondary in delta.

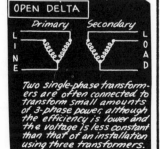

OPEN DELTA

Primary   Secondary

Two single-phase transformers are often connected to transform small amounts of 3-phase power, although the efficiency is lower and the voltage is less constant than that of an installation using three transformers.

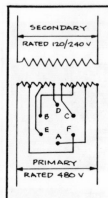

VOLTAGE TAPS *are often provided to enable the transformer to deliver rated voltage even though supply voltage may be below the rated value. In the example shown, connection may be made between any adjacent two of the six contacts by turning an external rotary switch. The table shows the switch connections used to maintain rated secondary voltage at various input voltages. Note that each successive tap accommodates a voltage reduction of 2½%. Taps are changed only with transformer deenergized.*

PERCENT RATED VOLTAGE	INPUT VOLTS	SWITCH CONNECTIONS
100 %	480 V	C TO D
97.5	468	B TO D
95	456	B TO E
92.5	444	A TO E
90	432	A TO F

000-volt services to handle the heavy electrical loads of modern buildings and widespread popularity of the 480Y/277-volt distribution system are the two forces behind the extensive application of transformers. In the former case, transformers are used to step the high voltage service or distribution down to a lower distribution and/or utilization voltage.

In the latter case, transformers are used to obtain the 120-volt energy required for incandescent lighting and receptacle outlets.

Various names are assigned to types of transformers according to their common use in electrical systems, as follows:

1. General-purpose transformers are the dry-type, air-cooled units, with primaries rated not over 600 volts, and used for local step-down from a distribution voltage to a utilization voltage level. Also called "power and light transformers" or simply "lighting transformers", they include wall-mounting, ceiling suspended and floor-standing units.

2. Load-center transformers include both dry-type and liquid-filled units, with their primaries usually rated from 2400 volts up to 15,000 volts. These units are used both indoors and outdoors to step from a primary voltage level down to a secondary voltage distribution and/or utilization level. Such transformers may be used either separately—in conjunction with separate protective and switching devices on the primary and secondary sides —or they may be combined with all necessary switching and protection in packaged unit substations, also called load-center substations.

3. Distribution transformers are liquid-cooled—oil or non-flammable askarel—units generally used outdoors on poles or on platforms. They include both single-phase and three-phase units, with primary ratings from 480 volts up to 15,000 volts.

4. Substation transformers are generally outdoor units of the oil or askarel type, with primary ratings in the high voltage range up to 67,000 volts or even higher. They are intended for use in utility distribution substations and outdoor industrial substations. Over 67 kv, transformers are commonly called "power transformers."

Transformer Selection

Because of their great importance to the successful operation of modern systems and because of their relatively high unit cost as part of an overall system, selection of transformers should be made with extreme care and careful attention to ratings and specifications. Consideration of basic characteristics should be made as follows:

Primary and secondary voltage ratings. Basically, the primary and secondary voltage ratings of a transformer must match the voltage levels established by circuit design. The rating of the primary winding must match the voltage of the supply circuit and the rating of the secondary winding must conform to the voltage requirements of the load circuit fed by the transformer. Transformers are made with a wide range of standard primary and secondary voltage ratings to meet the demands of standard system configuration. And units can always be supplied on demand to meet any levels and configurations of voltages.

As discussed previously, transformers can be obtained with various taps on the windings to adjust the voltage rating of the unit to the voltage rating of the supply circuit. These taps provide close application of transformers to practical systems in which voltage drop produces varying levels of voltage to a transformer depending upon the distance from the transformer back to the source of voltage of the primary circuit.

On general-purpose transformers, the primary winding is commonly provided with tap connections which afford about a 10% variation in the input voltage rating of the transformer. As an example, a transformer with a nominal primary voltage rating of 480 volts may be equipped with four 2½% taps below normal rated voltage. This would mean that the lowest tap would provide for connection of the primary winding to a supply circuit rated at about 432 volts (4 x 2½% x 480 volts=48 volts below 480 volts,

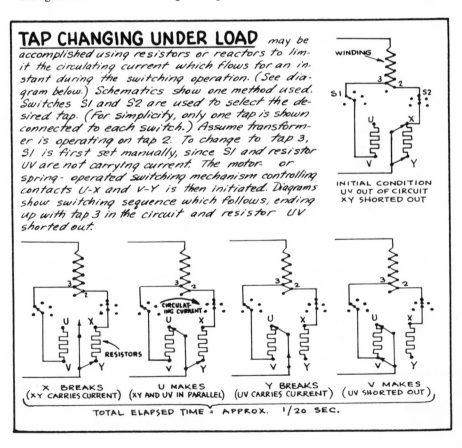

TAP CHANGING UNDER LOAD may be

accomplished using resistors or reactors to limit the circulating current which flows for an instant during the switching operation. (See diagram below.) Schematics show one method used. Switches S1 and S2 are used to select the desired tap. (For simplicity, only one tap is shown connected to each switch.) Assume transformer is operating on tap 2. To change to tap 3, S1 is first set manually, since S1 and resistor UV are not carrying current. The motor- or spring-operated switching mechanism controlling contacts U-X and V-Y is then initiated. Diagrams show switching sequence which follows, ending up with tap 3 in the circuit and resistor UV shorted out.

WINDING

INITIAL CONDITION
UV OUT OF CIRCUIT
XY SHORTED OUT

CIRCULAT-
ING CURRENT

RESISTORS

X BREAKS
(XY CARRIES CURRENT)

U MAKES
(XY AND UV IN PARALLEL)

Y BREAKS
(UV CARRIES CURRENT)

V MAKES
(UV SHORTED OUT)

TOTAL ELAPSED TIME ≈ APPROX. 1/20 SEC.

or 480 minus 48). Thus, connection of the 432-volt supply to those taps will produce the rated secondary voltage of 120/240 volts.

If the 432-volt supply were connected to the 480-volt terminals or to any of the other taps, the secondary voltage would be lower than the rated 120/240. But if the supply circuit were rated at, say 468 volts, connection of the supply would have to be made to the first tap below normal (1 x 2½% x 480 volts=12volts, which is then subtracted from 480) to obtain the correct rated secondary voltage.

Transformer taps change the value of voltage transformation by changing the ratio of primary turns to secondary turns. And because transformers are designed with their taps to handle the higher primary current at reduced primary voltage, the kva rating of a transformer is not changed by changing taps.

The kva rating of a transformer winding is established from the maximum allowable current rating of the winding and the voltage rating of the winding. Current x voltage x 1000 equals kva, for a single-phase winding. For the given voltage rating, the ampere rating is a maximum value set by the impedance of the winding ($I=E/Z$) and the design capability of the winding to dissipate heat.

Conditions of use must not place a higher current on either primary or secondary winding. If the voltage input to a given transformer winding is, say, half the rated voltage, the maximum kva rating of the transformer is thereby also reduced to half the normal value because the current rating cannot be increased (doubled) to keep the kva constant. If overvoltage is applied to a given transformer winding, the current will increase above rated value, producing overheating with resultant shortening of transformer life.

Two important definitions with respect to transformer ratings are as follows:

Rated Secondary Voltage of a Constant-Potential Transformer is the voltage at which the transformer secondary is designed to deliver rated kva.

Rated Secondary Current of a Constant-Potential Transformer is the secondary current obtained by dividing the rated kva by the rated secondary voltage.

EFFECTIVE selection of a transformer for any application involves consideration of the "impedance" of the transformer. This "impedance" of a transformer is the opposition which the transformer presents to the transfer of power through the unit—from primary to secondary. Although a transformer is a device for transferring power from one voltage level to another and is not a utilization device which is intended to consume power, the fact that it has windings with resistance and reactance means that it will offer some opposition to current flow from its secondary into a load or into a short circuit or other fault on the secondary side.

Every transformer has impedance and it is commonly expressed as a percentage of primary voltage. The percent impedance is the percent of normal rated primary voltage which must be applied to the primary winding to make rated load current flow with the secondary of the transformer short-circuited. Of course, such a definition sets up only a condition of test to determine the transformer impedance and does not suggest application in any such manner.

As an example of the meaning of impedance, if a 480-volt to 120-volt transformer is marked to have an impedance of, say, 5%, this means that full-load rated current can be made to flow by applying 5% of 480 volts or 24 volts to the primary winding. From such information, it is readily determined that full voltage of 480 volts applied to the primary will cause 20 times rated current (100% divided

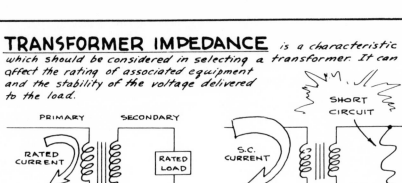

TRANSFORMER IMPEDANCE is a characteristic

which should be considered in selecting a transformer. It can affect the rating of associated equipment and the stability of the voltage delivered to the load.

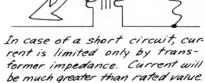

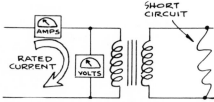

Rated current flows when transformer feeds rated load.

In case of a short circuit, current is limited only by transformer impedance. Current will be much greater than rated value.

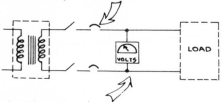

High transformer impedance means lower short-circuit current and lower interrupting capacity requirements for protective devices.

To reduce the short-circuit current to rated current, the impressed voltage would have to be reduced to a small fraction of its rated value. This fraction, expressed as a percentage of the primary voltage, is known as the IMPEDANCE VOLTAGE of the transformer.

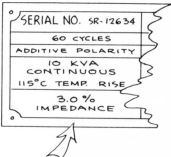

On the other hand, low transformer impedance means a more constant voltage available under varying load conditions.

Transformer selection must therefore be a compromise to best suit each individual application.

Tests made by the manufacturer determine this percentage for each transformer. It appears on the nameplate as "% IMPEDANCE."

by 5% equals 20) to flow when the secondary winding is short circuited.

For a given kva rating, the lower the percent impedance of a transformer, the higher will be the short-circuit current which the transformer can deliver into a short circuit on its secondary terminals. Take two transformers, each rated at 500 kva. The rated secondary load current is the same for both transformers. But if one of these transformers is rated at 10% impedance, it can supply 100%/10% or 10 times rated secondary current into a short circuit on its secondary terminals. And if the other 500 kva transformer has an impedance of 2%, it can feed 100%/2% or 50 times rated secondary current into a short on the secondary. As a result of the difference in impedance of the two transformers, the latter transformer can deliver five times as much short circuit current as the first one, even though they both have the same kva rating and ability to supply normal load current. Impedance values of general-purpose transformers generally run from 3% to 6%.

Another consideration in transformer application is that of voltage regulation, and transformer impedance is related to it. Voltage regulation is a measure of how the secondary voltage of a transformer varies as the secondary load current varies from zero to its full-load value, with the primary voltage held constant. Regulation is expressed as a percentage— no-load voltage minus full-load voltage divided by full-load voltage times 100%.

As an example of voltage regulation, if a transformer has a no-load secondary voltage of 240 volts and the voltage drops to 220 volts when the transformer is supplying its full-load rated current to a load, the regulation percentage is: (240 — 220) ÷ 220 x 100%=9%.

That indicates that 9% of the secondary voltage is being dropped across the internal impedance of the transformer. It is therefore obvious that the higher the impedance of a transformer, the greater will be the drop from no-load to full-load and the higher will be the percent regulation. It is generally desirable to keep regulation as low as possible to minimize variation in voltage as load-current demands vary. Typical regulation values are between 2% and 4%.

Low regulation percentage becomes very important when the transformer supplies varying load demands for utilization equipment which is sensitive to voltage fluctuations. However, when the load current on the trans-

former is relatively constant, taps on the transformer or primary supply conditions can be adjusted to provide a fixed compensation for voltage drop in the transformer.

Use of a transformer with higher than normal impedance and regulation may often offer advantage because it reduces the amount of short-circuit current that can flow. As a result, circuit breakers, fuses and other protective devices used on the secondary side of the transformer will not have to be rated for as high a current interrupting ability as they would have to be if a transformer of lower impedance were used.

KVA Rating

The kva rating of a transformer sets the maximum load which it can deliver for a specified time, at rated secondary voltage and frequency, without exceeding a specified temperature rise—based on the insulation life and ambient temperature. The load which a transformer can handle without deterioration of the insulation may be more or less than the rated output, depending upon ambient temperature and the time-duration of loading. A transformer may be loaded above rated kva only in accordance with prescribed temperature testing or short-time loading data from the manufacturer. The load on self-cooled transformers can be increased by the addition of fans which increase heat dissipation and permit higher loading before the temperature limit of safe operation is reached.

Selection of a transformer involves choosing a unit with kva rating at least equal to the kva of load. But spare capacity in transformers should always be considered for load growth.

Again, rated kva is based on drawing maximum rated current at maximum rated voltage. But rated kva is not a constant which permits operation at higher current if the voltage applied is lower or operation at higher voltage if the current is lower.

A number of rules of thumb are

available for selection of a transformer to supply motor loads. One general rule says: provide 1 kva of transformer capacity for each hp of motor load. Another rule recommends 1¼ kva per hp of motor load for motors rated 5 hp and above, with 1 kva per hp for smaller motors and motors rated over 50 hp.

Extreme accuracy can be achieved in matching transformer capacity to a given motor load if the nameplate current ratings of the motors are added together then multiplied by the circuit voltage and the product divided by 1000 to obtain minimum required kva. Usually it is necessary to select the standard transformer rating above the required kva.

Instrument Transformers

Direct connection between measuring instruments and high voltage circuits is not possible due to the delicate nature of the instruments which would be damaged by high voltages. Other factors also make it undesirable to use meters directly on high voltage.

A "potential transformer" is an instrument transformer used to connect a voltmeter to a high voltage ac circuit in which it is desired to measure the voltage. These transformers are similar to the constant-potential transformers used in power circuits, except that the power rating of the potential transformer is always low (50 to 500 watts). Potential transformers are available in the dry type and the oil-immersed type, depending upon their maximum voltage rating. Their primaries are rated for voltages like 35,000 volts; their secondaries are rated at 115-120 volts.

A "current transformer" is another type of instrument transformer used to connect ammeters and current coils of instruments into high voltage lines. One type of current transformer is constructed so that the cable in which current is to be measured can pass directly through a hole in the transformer assembly. The cable thereby becomes the primary of the transformer, even though only a short length of it passes through the assembly The secondary consists of a winding of many turns within the transformer assembly, and the measuring instrument is connected to the secondary. In this type of current transformer, used for measuring high currents in large conductors, there are no connections between the transformer and the large conductor. In current transformers of smaller rating, a primary winding within the transformer housing is often connected in series in the line in which the current is to be measured.

Raceways

ACCORDING to the National Electrical Code, a raceway is any channel used for holding wires, cables or busbars, which is designed and used solely for this purpose. A raceway may be constructed of metal or insulating material. The term raceway includes: rigid metal conduit, flexible metal conduit, electrical metallic tubing, underfloor raceways, cellular metal floor raceways, surface metal raceways, wireways, busways and auxiliary gutters. Each type of raceway has particular construction characteristics and application advantages. Although busways are classified as raceways, they contain electrical conductors as integral parts of their construction and constitute a separate technology.

Raceways are used to enclose wires and cables for several reasons. First, raceways provide mechanical protection for the wires and cables. They prevent accidental damage to the insulation and conductors, assuring long life and reliability of the wiring system. They also protect the wiring against harmful atmospheres and tampering. Another function of raceways is to provide electrical safety to both persons and property. Metallic raceway systems are bonded together and grounded to protect against electrical fault conditions and such hazards as lightning. In the event of fault in the wiring system or in the equipment or load devices used in the system, grounded metallic conduit will carry the current to ground, operating the system's protective devices—fuses or circuit breakers. Non-metallic raceways provide insulation and protect against exposure to dangerous voltages due to fault conditions.

Rigid Metal Conduit

The best known and most widely used raceway for industrial and com-

RIGID METAL CONDUIT

provides

- mechanical protection for conductors
- safety from electrical shock and fire
- convenient passage for conductors
- means for a continuous ground

To insure these benefits,
the National Electrical Code contains specific requirements governing the use of rigid metal conduit.

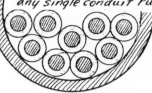

The National Electrical Code also governs the number of conductors permitted in any single conduit run.

CONDUIT FITTINGS of many types are available for changing direction of conduit run, aiding conductor pulling, making taps to other conduit runs, making splices, and connecting outlet equipment.

Cover

THREADED COUPLINGS are used to join successive lengths of conduit.

Quarter-bends

A maximum of four quarter-bends or equivalent is permitted in any one conduit run between fittings or outlets.

Quarter-bends

LOCKNUTS fasten conduit to box securely

BENDING is accomplished by means of hand bending tools or hydraulic benders.

Minimum radius

To insure against injury to conductors in pulling due to reduced internal conduit diameter, bends are required to have a radius at least 6 times the internal diameter of the conduit.

BUSHINGS are required where a conduit enters a box, to protect conductors from abrasion.

Both ends must be reamed to remove rough edges for protection of the conductors.

Manufactured in standard 10-ft lengths

Tapered threads insure tightness of fit between conduit and couplings, fittings, etc.

mercial wiring systems is rigid metal conduit. This type of raceway is made in the same sizes as standard pipe. The National Electrical Code requires that rigid conduit be made in lengths of 10 feet, including coupling. A length of rigid conduit is supplied with a coupling on one end and a special cap on the other end to protect the threaded end against accidental damage due to handling.

Rigid metal conduit may be made of steel, copper alloy or aluminum, according to processes which give the finished conduit the necessary characteristics for electrical application. Each threaded length of rigid conduit is given a finish which protects the conduit against weather and other atmospheric conditions. Steel conduit may be galvanized—either by electro-galvanizing or hot-dip galvanizing—or sherardized. These finishing treatments are based on the desirability of coating conduit with zinc. From tests and experience, zinc has been established as a highly durable coating which protects steel against atmospheric conditions, both indoors and outdoors. Application of zinc to rigid metal conduit by the processes of galvanizing or sherardizing is followed by an inside coating of varnish, lacquer or enamel to readily distinguish the conduit from other types of pipe used for non-electrical applications. This inside finish provides a smooth surface which greatly facilitates fishing and pulling of wires in the conduit.

Enameled rigid metal conduit is another type of coated electrical conduit. The enamel coating is applied by dipping and is baked to provide a tough, smooth finish. Enamel-coated conduit has its threaded ends protected during the enameling process. It may be used only indoors in dry locations.

Electrical code requirements for rigid steel conduit establish the minimum conditions for application and installation. Typical provisions of the code are as follows:

• Rigid metal conduit may be used in any type of building or occupancy under any types of atmospheric conditions, except that steel conduit protected from corrosion only by enamel may only be used indoors in dry non-corrosive atmospheres.

• Conduit and fittings exposed to severe corrosive atmospheres must have construction and finish suitable for the conditions and capable of resisting the corrosive influences. Occupancies where severe corrosive conditions commonly exist are: meat-packing plants, tanneries, hide cellars fertilizer rooms, some chemical plants, metal refineries, pulp and paper mills.

• Conduit used in cinder fill must be located 18 inches or more under the fill or must be protected on all sides by a layer of non-cinder concrete at least 2 inches thick, unless the conduit is corrosion-resistant and suitable for use in or under cinder fill where subject to permanent moisture.

• Installation of conduit in wet locations and places where walls are washed frequently must be carefully made to prevent water from entering the conduit system.

• The minimum allowable size of conduit is set at the nominal ½-inch trade size, with the exception of conduit used for certain underplaster extensions of an electrical system and conduit used to enclose the leads to an individual motor.

• A run of conduit between two outlets, two fittings or an outlet and a fitting may not contain more than the equivalent of four quarter bends (a total of 360 degrees).

IN ADDITION to rigid metal conduit, another common type of conduit used in modern electrical wiring systems is electrical metallic tubing, known also as "EMT" or "thin-

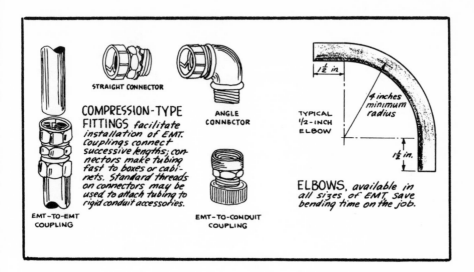

STRAIGHT CONNECTOR

COMPRESSION-TYPE FITTINGS *facilitate installation of EMT. Couplings connect successive lengths; connectors make tubing fast to boxes or cabinets. Standard threads on connectors may be used to attach tubing to rigid conduit accessories.*

ANGLE CONNECTOR

EMT-TO-EMT COUPLING

EMT-TO-CONDUIT COUPLING

TYPICAL ½-INCH ELBOW

1½ in

4 inches minimum radius

1½ in.

ELBOWS, *available in all sizes of EMT, save bending time on the job.*

wall conduit." This type of conduit is very much like rigid metal conduit except that the wall of the conduit is much thinner than that of rigid conduit. The National Electrical Code requires that electrical metallic tubing and the elbows and bends used with the tubing have a circular cross-section, that it have a finish or treatment of its outer surface to provide ready and permanent means of distinguishing it from rigid conduit when installed, and that threads used to couple tubing be so designed as to prevent bending of the tubing at any part of the thread.

Electrical metallic tubing is generally made of a special steel selected for its strength and bending characteristics, suiting it to use as a raceway system for electrical conductors. EMT has an outside zinc coating and an inside corrosion-resisting coating of either zinc or enamel which also provides a smooth finish for pulling wires in the tubing. EMT is extensively used with threadless couplings and fittings of the compression type.

Application

In general, the application and installation of electrical metallic tubing is regulated by the safety require-ments of the National Electrical Code. Typical provisions of the code are as follows:

• Electrical metallic tubing may be used for both exposed and concealed work. The tubing must have protection against corrosion, other than just enamel. Electrical metallic tubing **may not be used** under certain conditions: (1) where the tubing will be exposed to possible severe mechanical damage either during installation or afterwards, (2) where the tubing is embedded in cinder concrete or fill where subject to permanent moisture. Tubing may be used, however, if it is protected on all sides by at least 2 inches of non-cinder concrete or if it is at least 18 inches under the fill; (3) where the location of the installation is classified as hazardous and the National Electrical Code prohibits the use of tubing.

Hazardous locations are divided into three classes: Class I—locations in which flammable gases or vapors are or may be present in the air in sufficient quantities to produce explosive or ignitible mixtures; Class II—locations which are hazardous because of the presence of combustible dust; Class III—locations which are hazardous because of the presence of easily ignitible fibers or flyings, but in which such fibers and flyings are not

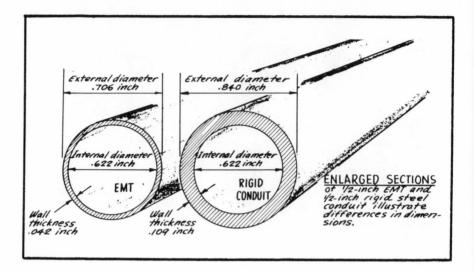

Enlarged sections of 1/2-inch EMT and 1/2-inch rigid steel conduit illustrate differences in dimensions.

likely to be in suspension in the air in quantities sufficient to produce ignitible mixtures. Further breakdown of hazardous locations is made according to the severity of the hazardous conditions within each class: Division 1 locations represent the most severe hazard; Division 2 locations are less severely hazardous. Code requirements on the use of electrical metallic tubing in such locations are as follows:

In Class I, Division 1 locations— only rigid conduit may be used, EMT **may not** be used.

In Class I, Division 1 locations— either rigid conduit of threaded tubing may be used.

In Class I, Division 2 locations— EMT **may not** be used.

In Class II, Division 2 locations— either rigid conduit or EMT may be used.

In Class III, Division 1 and Division 2 locations—EMT **may not** be used.

● Thinwall conduit may not be used for general wiring systems of more than 600 volts between conductors. It may be used, however, for out-door sign and outline circuits of more than 600 volts.

● In general, tubing should not be used where it would be exposed to corrosive vapors. If it is used in locations in which corrosive atmospheres prevail—meatpacking plants, tanneries, fertilizer rooms, chemical plants, refineries, etc.—the tubing and its fittings must be of corrosion-resistant construction suitable for the application.

● In wet locations or places which are washed or scrubbed down regularly —dairies, laundries, canneries, etc.— the tubing system and its fittings must provide a watertight system. Supports, clamps, bolts, etc., used with the system must be corrosion-resistant.

● The smallest size of electrical metallic tubing for general use is the ½-inch electrical trade size. Exceptions to this are made to allow use of smaller size tubing for underplaster extensions and for enclosing motor leads. The maximum size of tubing is 2-inch electrical trade size.

● The number of conductors allowed in a single run of tubing is the same as it would be for rigid conduit of the same electrical trade size. The same tables from the code apply for conductors in EMT. The allowance of greater conduit fill for rewiring also applies to EMT. The inside diameters of tubing and rigid conduit of the same electrical trade size are equal. Of course, EMT has a smaller outside diameter than that of rigid conduit of the same trade size.

• Tubing shall not be coupled together or connected to boxes, fittings or cabinets by means of threads in the wall of the tubing, except with approved fittings. Threadless couplings and connectors must be made up tight and must be watertight if used on tubing buried in masonry, concrete, fill or in wet places.

• Bends in EMT must be made without damaging the tubing or reducing its internal diameter. Not more than the equivalent of four quarter bends may be made in a run of tubing between outlet and outlet, between fitting and fitting or between outlet and fitting.

Article 510 of the code covers application of electrical metallic tubing in specific occupancies:

Commercial garages—Threaded rigid conduit or EMT with threaded fittings must be used in hazardous areas.

Gasoline stations—Only threaded rigid conduit may be used for pumping islands. Wiring below ground within 20 feet of pumps must be in rigid conduit. Wiring in this area within 18 inches above ground must be either threaded rigid or EMT.

Bulk storage plants—EMT may be used for wiring within 25 feet of loading platforms or tanks above ground, but rigid conduit must be used in rooms with tanks or pumps. Other cases are given in the code.

F LEXIBLE METAL conduit and surface metal raceway are used to supplement rigid metal conduit and electrical metallic conduit for general branch circuit and feeder wiring. Each of these types of raceways has particular application advantages and limitations in wiring for new construction and electrical modernization.

Flexible Metal Conduit

The use of flexible metal conduit, commonly called "Greenfield" or "flex," for general interior wiring com-

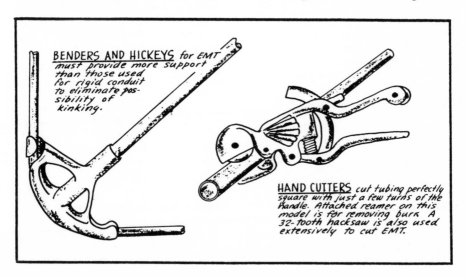

BENDERS AND HICKEYS for EMT must provide more support than those used for rigid conduit to eliminate possibility of kinking.

HAND CUTTERS cut tubing perfectly square with just a few turns of the handle. Attached reamer on this model is for removing burr. A 32-tooth hacksaw is also used extensively to cut EMT.

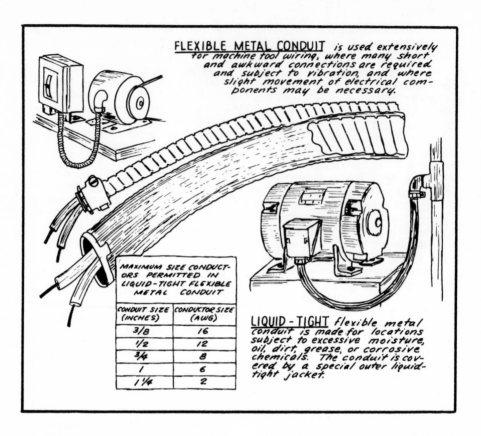

FLEXIBLE METAL CONDUIT is used extensively for machine tool wiring, where many short and awkward connections are required and subject to vibration, and where slight movement of electrical components may be necessary.

MAXIMUM SIZE CONDUCTORS PERMITTED IN LIQUID-TIGHT FLEXIBLE METAL CONDUIT	
CONDUIT SIZE (INCHES)	CONDUCTOR SIZE (AWG)
3/8	16
1/2	12
3/4	8
1	6
1 1/4	2

LIQUID-TIGHT flexible metal conduit is made for locations subject to excessive moisture, oil, dirt, grease, or corrosive chemicals. The conduit is covered by a special outer liquid-tight jacket.

bines the requirements for rigid metal conduit and armored cable. The method is similar to that of rigid conduit except that a flexible steel conduit is used, and the installation techniques are similar to those used for armored cable. Unlike armored cable, however, flexible metal conduit is only a raceway and contains no conductors. Runs of flexible metal conduit are installed from equipment enclosures to boxes or other equipment enclosures and then the wires are pulled in, from box to box, etc.

Flexible metal conduit is used for many particular wiring jobs. It is much superior to rigid conduit for runs through spaces containing many obstructions, such as water pipes, structural members, air ducts, etc.

In such areas, it is often impossible to run rigid conduit by making all the bends and turns necessary to avoid the obstructions. Installation of flexible conduit is quicker and easier than that of rigid conduit. And it comes in long continuous lengths up to 250 feet, whereas rigid conduit is available only in short lengths.

A common use of flexible conduit: to connect lighting fixtures above suspended ceilings. In many areas where lighting fixtures may have to be moved to suit changes in work performed, a common technique is to use a grid system of rigid or thinwall conduit with outlet boxes at all intersections, connecting with flex to the recessed fixtures.

Another common use of flexible conduit is in connecting rigid conduit to motors or machines. The flexible conduit in such cases absorbs vibrations which would set up noise and loosen rigid conduit if it were con-

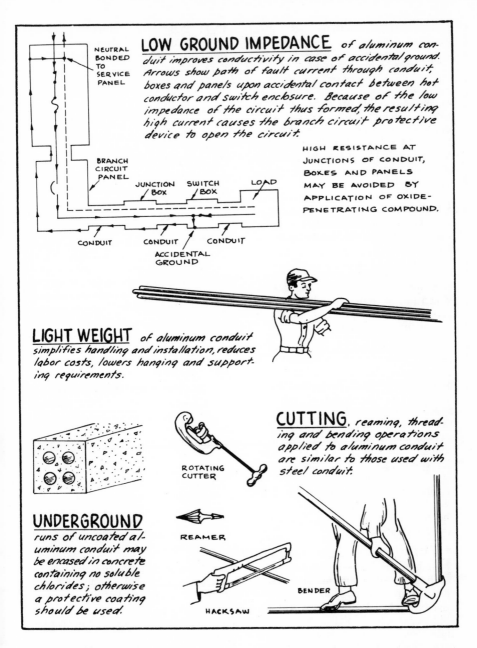

LOW GROUND IMPEDANCE of aluminum conduit improves conductivity in case of accidental ground. Arrows show path of fault current through conduit, boxes and panels upon accidental contact between hot conductor and switch enclosure. Because of the low impedance of the circuit thus formed, the resulting high current causes the branch circuit protective device to open the circuit.

NEUTRAL BONDED TO SERVICE PANEL

BRANCH CIRCUIT PANEL

JUNCTION BOX

SWITCH BOX

LOAD

CONDUIT   CONDUIT   CONDUIT

ACCIDENTAL GROUND

HIGH RESISTANCE AT JUNCTIONS OF CONDUIT, BOXES AND PANELS MAY BE AVOIDED BY APPLICATION OF OXIDE-PENETRATING COMPOUND.

LIGHT WEIGHT of aluminum conduit simplifies handling and installation, reduces labor costs, lowers hanging and supporting requirements.

CUTTING, reaming, threading and bending operations applied to aluminum conduit are similar to those used with steel conduit.

ROTATING CUTTER

UNDERGROUND runs of uncoated aluminum conduit may be encased in concrete containing no soluble chlorides; otherwise a protective coating should be used.

REAMER

HACKSAW

BENDER

nected directly to the vibrating machines.

Limitations on the use of flexible metal conduit are clearly set forth in the National Electrical Code. They establish that flexible metal conduit **shall not be used—**

1. in wet locations, unless conductors are lead-covered or other type approved for the conditions;

2. in hoistways, except for limited use in hoistways and escalator wellways between risers and limit switches, interlocks, operating buttons and similar devices;

3. in storage battery rooms;

4. in any hazardous location, except as specially allowed in the National Electrical Code under "Wiring Methods" when flexible connections are required for Class II (presence of combustible dusts) or Class III (presence of easily ignitible fibers or flyings) locations; or

5. where rubber-covered conductors are exposed to oil, gasoline or other materials having a deteriorating effect on rubber.

Liquid-tight flexible metal conduit is a special form of flexible conduit which has an outer liquid-tight jacket and is used with terminal fittings designed for the conduit.

This type of flexible conduit is not used as general purpose raceway. The code restricts its use to connection of motors or portable equipment where flexibility of connection is required. Flexible metal conduit of the liquid-tight type **may not be used** where subject to mechanical injury, where in contact with rapidly moving parts, under conditions such that its temperature is above 60 C (140 F), or in any hazardous location other than those allowed under "Wiring Methods" in the code sections covering Class I, II and III locations.

IN RECENT years, aluminum rigid conduit has gained widespread acceptance as an alternative to rigid steel conduit. Aluminum conduit is covered in Article 346 of the National Electrical Code—which also applies to rigid steel conduit. This article is entitled "Rigid Metal Conduit."

Characteristics

Available in a full range of standard sizes from ½-in up to 6-in diameter, including ells and couplings, aluminum conduit offers many advantages for electrical system use:

Light Weight — Aluminum conduit weighs only one-third as much as the same sizes of rigid steel conduit. This light weight lowers the labor requirements associated with handling and installing the conduit and also reduces the requirements for hanging or supporting the conduit in its installed position. Substantial economy is thus offered.

Corrosion Resistance—The natural finish of aluminum oxide on the surface of aluminum conduit provides a high degree of corrosion resistance. The conduit is resistant to weather, cannot rust and does not need painting. It also is resistant to many chemical atmospheres and has long found successful application in such corrosive atmosphere as sewerage disposal plants and a wide variety of chemical plants.

Reduced Voltage Drop—Because of the non-magnetic character of aluminum conduit the inductive reactance of circuit conductors run in the conduit is lower than it would be in steel conduit. This minimized inductive reactance reduces the voltage drop in the contained circuits, offering important improvement in voltage level and regulation of 60-cycle circuits. Particular advantage is provided for voltage-drop reduction in high-frequency circuits which have higher reactance in proportion to the frequency increase over 60 cycles. The non-magnetic character often makes aluminum conduit the only possible choice for high frequency circuits.

Low Ground Impedance—Because of the better electrical conductivity of aluminum compared with steel, a system of aluminum conduits offers lower impedance in the equipment grounding circuit.

The equipment grounding circuit consists of all of the metallic enclosures for electrical conductors and equipment-panel cabinets, switch enclosures, motor housings, conduit, raceways, etc. These enclosures do not normally carry current. But in the event of an insulation failure in the contained conductors or equipment, an electric potential could be placed on some such metallic enclosure.

The purpose of the equipment grounding system is to assure that all such metallic enclosures are firmly connected to each other and that the entire interconnected system is connected to the system grounding electrode—the water pipe, in most cases. With such connection to ground, no dangerous potential can appear on enclosures.

A separate electrical conductor can be used throughout an electrical system to tie all enclosures together and to ground. But the National Electrical Code permits the use of metal conduit to serve the purpose of grounding equipment housings and enclosures. Of course, the economy of such practice lies in eliminating the need for a separate grounding conductor by using the conduit for the dual purposes of mechanical protection for wires and an equipment ground path.

The lower electrical resistance and reactance of aluminum conduit permits higher level of current flow on faults to provide greater assurance of the operation of overcurrent devices in today's grounded electrical systems. This will minimize conditions in which the level of fault current is not high enough to operate the overcurrent device and the system of metal enclosures has a hazardous potential above ground on it. That is, the fuse or circuit breaker will not draw enough current to open the circuit and clear the fault.

To realize the foregoing advantage of aluminum conduit as an equipment grounding path, it is important to eliminate the potentially high resistances at conduit terminals and couplings due to the aluminum oxide on the surface of the conduit. This aluminum oxide has very high electrical resistance and can produce objectionably high resistance of the grounding path. Use of the oxide-penetrating compound used for aluminum conductors can eliminate high-resist-

ance connections. The compound—made of small metallic particles in a special grease—breaks down the oxide resistance at threaded couplings and locknut-and-brushing terminals, thereby assuring the desired overall low resistance of the grounding path.

Installation

Installation techniques for aluminum conduit generally follow those used for rigid steel conduit, with some special techniques finding favor among electrical contractors. For cutting aluminum conduit, a rotating abrasive-wheelcutter has been found to give quick, burr-free cuts, with no flaring-in of the end, ready to ream and thread. In the smaller sizes, a hacksaw offers quick cutting. As with steel conduit, a sharp cutting die is necessary for clean threading of conduit ends.

Although use of dissimilar metals—such as aluminum and steel—in contact with each other presents some possibility of galvanic corrosion when moisture is present, there has been much application of aluminum conduit with galvanized steel cabinets and boxes without any corrosion because aluminum and zinc are close together in the galvanic series. For the same reason, galvanized fittings can be used with aluminum conduit for dry locations indoor and for outdoor locations where there is low moisture, provided the zinc coating is intact.

Burial of aluminum conduit in concrete requires careful attention to the nature of the concrete. Over many years there have been trouble-free installations of aluminum conduit in normal concrete, that is, concrete which does not contain soluble chlorides—such as calcium chloride added as an "accelerator," coral-bearing aggregates, unwashed beach sand or sea water. When any of these conditions of soluble chlorides exist in the concrete, the aluminum conduit should be coated with a protective coating of bitumastic or asphalt-base paint, or a similar compound.

Aluminum conduit can be buried

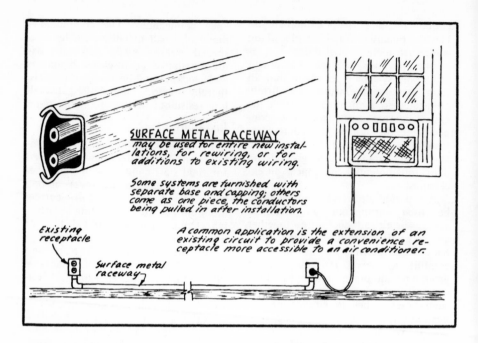

SURFACE METAL RACEWAY may be used for entire new installations, for rewiring, or for additions to existing wiring.

Some systems are furnished with separate base and capping; others come as one piece, the conductors being pulled in after installation.

Existing receptacle

Surface metal raceway

A common application is the extension of an existing circuit to provide a convenience receptacle more accessible to an air conditioner.

directly in the earth, but this is not recommended because some alkaline soils can corrode aluminum. Direct burial can also corrode steel conduit. Protective painting, taping or concrete encasement will prevent such corrosion.

Surface Metal Raceway

Surface metal raceway is a thin sheet-steel enclosure used to contain wiring, run along walls, ceiling or floors.

It may have a rectangular or flattened-oval cross section and may consist of two parts—a trough which is fastened to the surface of the wall or ceiling and in which the wires are placed and a capping which is snapped on over the wires in the trough—or may be simply a one-piece tubing through which wires must be pulled as they are in regular rigid or thin-wall conduit. Surface metal raceway systems are available with special fittings to allow for outlet or switch connections.

In general, surface metal raceway is used exposed in dry locations. **It may not be used:** where concealed, except that approved metal raceways may be used for underplaster extensions; where subject to possible severe mechanical injury unless approved for the purpose; where the voltage is 300 volts or more between conductors unless the metal has a thickness of not less than .040 inches; where subject to corrosive vapors; in hoistways; in storage battery rooms; or in any hazardous location.

The largest size conductor which may be used in surface metal raceway is No. 6. The number of conductors installed in any surface metal raceway **may not** be greater than the number for which the raceway is approved, but must never be greater than 10 conductors in a single raceway compartment.

Surface metal raceways, except multi-outlet assemblies, may be extended through dry walls, dry partitions and dry floors, if they pass through in unbroken lengths.

If combination surface metal raceways are used for both light power wiring and for signal or communication wiring, the different wiring systems should be run in separate compartments—identified by the use of sharply contrasting colors of the interior finish and by maintaining the same relative position of compartments throughout the raceway system on the premises.

IN MODERN electrical systems, raceway applications include widespread use of underfloor raceways, cellular metal floor raceway, multi-outlet assemblies and many types of wireways. Each of these types of raceways has particular application advantages and must be used in accordance with special code requirements.

Underfloor Raceways

Underfloor raceway systems are widely used in modern commercial, industrial and institutional buildings. Both metallic and non-metallic underfloor ducts are made for single, double or triple duct runs to provide separate channels for running power and light circuits, signal and control circuits and communication circuits.

Underfloor raceway systems consist of duct runs between special junction boxes which are part of the system, all set below the surface of the concrete floor slab. Connections of floor outlets can be made through inserts cast in the floors, directly into underfloor raceways or rigid conduit branches can be run from the junction boxes. With this system, floor outlets can be spotted at regular intervals or at selected locations, either for present use or covered for use at a later date.

The use of underfloor distribution meets the demand for many and conveniently located outlets in modern interiors, providing flexible and ready utilization of lighting units, business machines, appliances, telephones, intercom and signal devices. Such a system provides for changes in office layout or change in the activity pursued in any area. And underfloor raceway systems can be used with any type of concrete slab construction and with wood floors. Connection between an underfloor raceway system and wall-mounted panelboards is made with conduit runs down in the wall and into the slab to the junction box.

A variation on underfloor raceway is flush floor raceway which is set flush in the slab in office occupancies and covered with linoleum or similar floor covering.

Underfloor raceways are prohibited by the code:
(1) if subject to corrosive vapors, (2) in any hazardous location, (3) in commercial garages and (4) in storage-battery rooms. Raceways are approved for use with particular sizes of conductors, but No. 1/0 is the maximum conductor size allowed in any underfloor raceway. The sum of cross-sectional areas of conductors in a raceway must not exceed 40 per cent of interior cross-sectional area of the raceway, except in the case of armored cable or type NM (non-metallic sheathed) cable.

Cellular Floor Raceway

Cellular metal floor raceway systems consist of structural load-carrying floors made up of hollow or cellular steel construction plus electrical fittings and accessories required to provide access to the cells in the floor to use them as raceways for power and light circuits and for signal and communication wiring. Several such floor constructions are available for use as raceways. Typical accessories needed to convert the structural floor to use as raceway include: header ducts which have openings on top and bottom to afford passage of wires from the header to the floor cells; vertical cells to extend from header ducts to panelboards or cabinets; floor outlet devices; junction units for use where wires turn from headers into floor cells; and floor covering adapters for use on junction boxes.

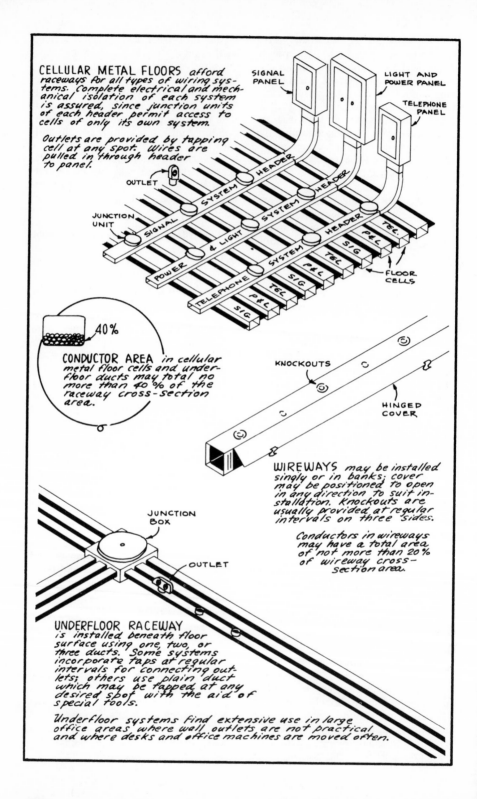

CELLULAR METAL FLOORS *afford raceways for all types of wiring systems. Complete electrical and mechanical isolation of each system is assured, since junction units of each header permit access to cells of only its own system.*

Outlets are provided by tapping cell at any spot. Wires are pulled in through header to panel.

SIGNAL PANEL

LIGHT AND POWER PANEL

TELEPHONE PANEL

OUTLET

JUNCTION UNIT

SIGNAL SYSTEM HEADER

POWER & LIGHT SYSTEM HEADER

TELEPHONE SYSTEM HEADER

SIG

P&L

TEL

FLOOR CELLS

40%

CONDUCTOR AREA *in cellular metal floor cells and underfloor ducts may total no more than 40% of the raceway cross-section area.*

KNOCKOUTS

HINGED COVER

WIREWAYS *may be installed singly or in banks; cover may be positioned to open in any direction to suit installation. Knockouts are usually provided at regular intervals on three sides.*

Conductors in wireways may have a total area of not more than 20% of wireway cross-section area.

JUNCTION BOX

OUTLET

UNDERFLOOR RACEWAY *is installed beneath floor surface using one, two, or three ducts. Some systems incorporate taps at regular intervals for connecting outlets; others use plain duct which may be tapped at any desired spot with the aid of special tools.*

Underfloor systems find extensive use in large office areas, where wall outlets are not practical and where desks and office machines are moved often.

Header ducts are installed at right angles to the floor cells, with junction units spaced to correspond to cell centerlines for wiring from headers to cells. Separation of different electrical systems—power and light, signals, etc.—is accomplished by using separate headers connected to panelboards or cabinets supplying the different systems. Then each header provides access to distinct cells accommodating only one of the different electrical systems. Outlets are made only from the floor cells, which serve as the actual raceway, and never from the header ducts.

Multi-Outlet Assembly

Multi-outlet assemblies are surface type raceway systems for branch circuit wiring, with convenience receptacle outlets spaced at fixed or adjustable positions along the length of the raceway. Such assemblies are made in several styles for wall mounting, mounting along baseboard, mounting in baseboard, etc. Such assemblies provide convenient use of electric power in residential, industrial and commercial areas requiring flexible use of plug-in appliances.

According to the National Electrical Code, multi-outlet assembly may be used in dry locations. **It may not be used:** concealed, except that it may be recessed; where subject to severe mechanical injury; and where general use of surface metal raceway is prohibited as described in the foregoing section. Multi-outlet assembly may not be extended through walls or floors.

Wireways

Wireways are sheet-metal troughs with hinged or removable covers for housing and protecting electrical wires and cables and in which conductors are laid in place after the wireway has been installed as a complete system. Wireways may be standard manufactured type or custom made for particular applications. They may be used only for exposed work in dry loca-

tions; **they may not be used** where subject to severe mechanical injury or corrosive vapor, in hoistways, in any hazardous location or in storage-battery rooms.

The largest size conductor allowed in any wireway is 500 MCM. No more than 30 conductors may be installed through a wireway at any cross-section, except in the case of conductors between a motor and its starter and used only for starting duty.

Wireways are widely used for enclosing feeders or mains when a large number of conductors must be located in one run, for enclosing splices and taps from feeders or subfeeders to transformers, switches, or branch circuits, for enclosing motor control and other control circuits.

TWO RACEWAY methods which have found increasing application over recent past years are cable trays and non-metallic conduit. Much discussion and controversy have ensued from these applications relating to the safety and adequacy of the products themselves and their installation techniques. To regulate the use of these products in a manner consistent with maximum safety in electrical work, the 1962 edition of the National Electrical Code presents, for the first time, specific requirements on both enclosures.

Cable Tray

Cable trays and racks are classified by the Code as "Continuous Rigid Cable Supports." This is presented as Article 318 in the Code and covers the two basic forms of this product: **1. expanded metal (mesh) trays;** and **2. cable racks** (which generally resemble ladder sections). These metal trays and racks are available in steel and aluminum with various methods of coupling sections together, such as splice plates, pins, bolts, and clamping channels. Standard accessory parts include angle, tee and offset sections to be used with straight lengths. The trays and racks are avail-

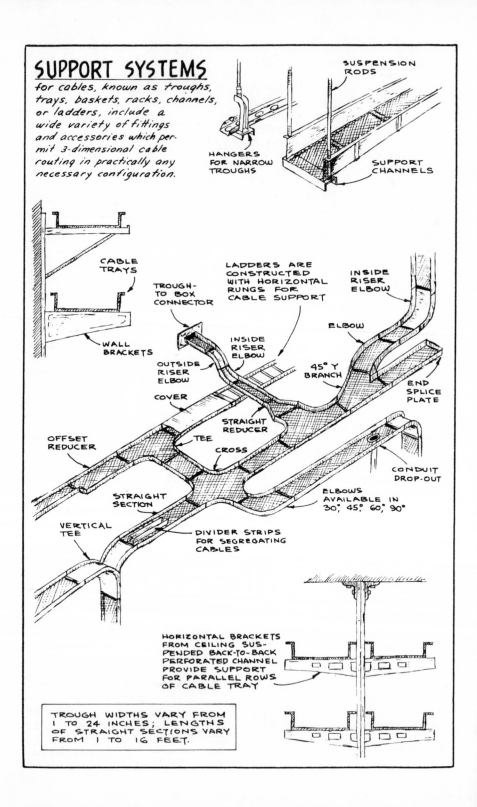

SUPPORT SYSTEMS

for cables, known as troughs, trays, baskets, racks, channels, or ladders, include a wide variety of fittings and accessories which permit 3-dimensional cable routing in practically any necessary configuration.

SUSPENSION RODS

HANGERS FOR NARROW TROUGHS

SUPPORT CHANNELS

CABLE TRAYS

TROUGH-TO BOX CONNECTOR

LADDERS ARE CONSTRUCTED WITH HORIZONTAL RUNGS FOR CABLE SUPPORT

INSIDE RISER ELBOW

WALL BRACKETS

INSIDE RISER ELBOW

ELBOW

OUTSIDE RISER ELBOW

COVER

45° Y BRANCH

END SPLICE PLATE

OFFSET REDUCER

STRAIGHT REDUCER

TEE

CROSS

CONDUIT DROP-OUT

STRAIGHT SECTION

ELBOWS AVAILABLE IN 30°, 45°, 60°, 90°

VERTICAL TEE

DIVIDER STRIPS FOR SEGREGATING CABLES

HORIZONTAL BRACKETS FROM CEILING SUS-PENDED BACK-TO-BACK PERFORATED CHANNEL PROVIDE SUPPORT FOR PARALLEL ROWS OF CABLE TRAY

TROUGH WIDTHS VARY FROM 1 TO 24 INCHES; LENGTHS OF STRAIGHT SECTIONS VARY FROM 1 TO 16 FEET.

able in various widths—6-in., 12-in., 18-in., etc.—to support varying numbers of cables.

Cable trays and racks are especially suited to feeder and long branch circuit runs to individual load devices. They particularly solve the problems associated with installing multiplicities of power and control cables in modern industrial and commercial electrical systems.

In open industrial applications, trays and racks have been selected for their ready accessibility to the wires installed in them, for the extreme flexibility offered for rerouting circuits to meet shifting loads or load centers. And the open, ventilated manner of running cables gives greater dissipation of heat generated in current-carrying cables, offering higher utilization of copper capacities, cooler operation of the cables and reduced possibility of heat damage to cable insulation.

In past years, the most common case of cable tray distribution found multiple-conductor, neoprene-jacketed cables strapped in the tray (or on the ladder rack). Such cable assemblies constituted a Code-recognized wiring method when the cable was type USE (underground service entrance cable) installed according to Articles 336 and 338 for interior wiring.

The Code now permits the use of tray or rack only as a support system for wiring methods already recognized as complete in themselves. This means tray or rack may be used to support metal-clad cable, aluminum-sheathed cable, mineral-insulated cable, non-metallic sheathed cable, UF cable, service entrance cable (as used for interior wiring) or any approved conduit or raceway with its own contained conductors.

Trays and racks are not considered to be "raceways" in the Code meaning of the word. They are not suitable for use with building wire as are conduit, EMT, flex, wireway, underfloor raceway and surface raceways—all of which come under the Code definition of "raceway" given in Article 100 of the Code.

Specific Code rules are:

• Continuous rigid cable supports may be ventilated or non-ventilated, but must be non-combustible.

• Not more than two layers of cables or conduits are permitted in a tray or rack. And if type MC metal-clad cable is used, only one layer is permitted.

• When a second layer of cable or raceway is installed in a tray or rack, a continuous, ventilated non-combustible separator to support the second layer must be placed in the tray at least one-inch above the first layer of cable or raceway.

• The vertical spacing between continuous rigid cable supports must be not less than 18 inches center to center.

• Tray or rack may be used exposed in wet or dry locations, in fire-resistive or non-combustible construction or other construction approved by the local inspector.

• Tray or rack may not be used in hazardous locations, in hoistways or where subject to severe physical damage.

• Current rating of conductors used in special cables in tray or rack must be derated in accordance with Section 318-8(e).

Rigid Non-Metallic Conduit

New Code Article 347 in the 1962 NE Code for the first time recognizes and covers application of conduits of non-metallic material which is resistant to moisture and chemical atmospheres. This includes: fiber, asbestos cement, soapstone, rigid polyvinyl chloride and high-density polyethylene for underground use and only rigid polyvinyl chloride (abbreviated, PVC) for use above ground.

Rigid non-metallic conduit may be used for circuits up to 600 volts (or over 600 volts for direct burial) under the following conditions:

1. For direct earth burial, not less than 24 inches below grade (and encased in two inches of concrete where the voltage exceeds 600 volts).

2. In locations subject to severe corrosive influences, PVC conduit may be used for above ground and in-building wiring in accordance with Section 300-5.

3. In cinder fill.

4. In wet locations, with the entire conduit system installed to prevent water from entering the conduit.

5. In concrete walls and floors.

Uses 2, 4 and 5 above cover non-metallic conduit as a wiring method in buildings for which only PVC is recognized.

Rigid non-metallic conduit must not be used: above ground outdoors, in hazardous locations, in the concealed spaces of combustible construction, for the support of fixtures or other equipment.

It must be noted that use of non-metallic conduit eliminates the equipment grounding function normally performed by metallic conduit and raceway through their solid inter-connection to boxes, cabinets, motor housings, lighting fixture enclosures, etc. With non-metallic conduit, a separate grounding conductor will generally be required, run in the conduit, to provide the interconnected network of grounding of all non-current-carrying metallic parts of electrical equipment. This extra grounding conductor will frequently require use of a larger size of conduit than would be required if metal conduit were used without the grounding wire. But for many applications, such as wet and corrosive locations, non-metallic conduit offers substantial advantage over other wiring methods.

Busways

ONE OF THE fastest growing applications in the electrification of modern commercial and industrial buildings is busway.

In the past, busway systems have been widely used for feeders, plug-in subfeeders and for plug-in and trolley type branch circuits in industrial occupancies. But the use of feeder and plug-in busway in large commercial and institutional buildings is only now coming into general acceptance.

It has been successfully used to carry power up to the different floors in high office buildings, and such buildings as convention halls and apartment houses are finding use for both horizontal and vertical busway power distribution circuits.

Engineering design of busway systems is often considerably less involved than design of wire and conduit distribution circuits.

Busways permit frequent changes in load layout, adapting easily to rearrangement of production line techniques in plants, without serious disruption of production.

Another very important advantage of busways systems is their controlled voltage drop characteristics which permit effective application to meet varying requirements for voltage stability. And the sizes in which busways are available permit the electrical designer to select busway which has sufficient spare carrying capacity to meet future electrical load growth.

Applications

Although busways have broad application potential, they are limited. The National Electrical Code limits their application to exposed work. This answers the question which is frequently raised about the use of busways for concealed work.

In at least one case on record, however, local code authorities permitted an exception to the rule. The case involved the use of concealed busway used along the back wall of a modern shopping center to supply service entrance equipments for different stores in the center. Of course, local Code authorities are empowered by the Code itself to rule on special applications involving unusual conditions.

The Code prohibits the use of busways where they might be subject to mechanical injury or corrosive vapors, in hoistways, in any hazardous location, in storage-battery rooms or outdoors or in damp locations unless specially approved for the purpose.

Specific approval, however, is given for their use as service entrance conductors. To secure effective mounting of busway, the Code requires that supports should be provided every five feet unless approved for supporting

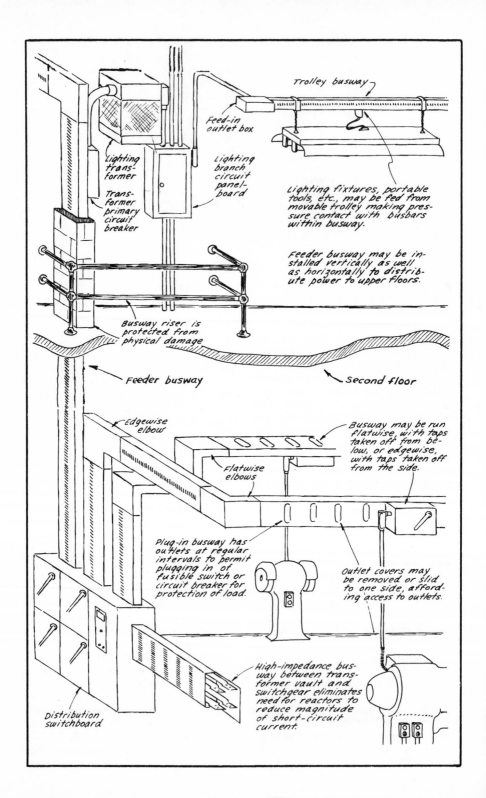

Trolley busway

Feed-in outlet box

Lighting branch circuit panel-board

Lighting transformer

Transformer primary circuit breaker

Lighting fixtures, portable tools, etc., may be fed from movable trolley making pressure contact with busbars within busway.

Feeder busway may be installed vertically as well as horizontally to distribute power to upper floors.

Busway riser is protected from physical damage

Feeder busway

Second floor

Edgewise elbow

Flatwise elbows

Busway may be run flatwise, with taps taken off from below, or edgewise, with taps taken off from the side.

Plug-in busway has outlets at regular intervals to permit plugging in of fusible switch or circuit breaker for protection of load.

Outlet covers may be removed or slid to one side, affording access to outlets.

High-impedance busway between transformer vault and switchgear eliminates need for reactors to reduce magnitude of short-circuit current.

Distribution switchboard

at greater intervals. In no case, however, may busway be supported at intervals greater than 10 feet.

Made up of copper or aluminum busbars mounted on insulated supports within a rigid steel housing, busway is available in a number of different types and sizes for use in distribution systems.

Feeder Busway

One of the most widely used types of busway is the **low-impedance type** used for feeder applications. This type of busway has low voltage drop characteristics due to close spacing of the busbar conductors and interleaved configuration of the different phase bars which greatly lowers the inductive reactance of the length of busway. It is used for all types of high capacity feeders and risers and is especially effective for use in high frequency systems.

Low-impedance busway is used for service runs from transformer vaults to switchboards, for feeders from switchboards to load center panels or to busway subfeeders, and for welder feeders.

Busway with aluminum conductors is recommended for use where sulfurdioxide atmospheres prevail. Standard feeder busways for inside use are made with ventilated housing to assure low operating temperature.

Other types of housing are available for outdoor use and for moist and other special atmospheres. Typical ratings on low-impedance busway range from 6,000-amps to 4,000-amps, for single-phase (2 or 3 poles) or 3-phase (3 or 4 poles), for 120/240 volts single-phase, for 120/208 volts or 480/277 volts 3-phase, 4-wire systems or for 3-phase, 3-wire systems up to 600 volts.

Plug-In Busway

Plug-in distribution busway systems offer extreme flexibility to industrial plants. Plug-in busway is generally similar to standard busway except that it has easily accessible plug-in openings along its length to permit tapping into the busbars to supply various loads through plug-in switches and circuit breakers.

This type of busway is available in ratings from 225- to 1,500-amps for secondary feeder systems rated 600 volts and below. Plug-in busway can be run from a switchboard or tapped from a run of feeder busway to carry power to closely spaced machines and other loads. Branch circuits to the motors can be tapped from the plug-in busway.

Branch Busway

Busways for branch circuits are made in 2-, 3- and 4-pole types of plug-in and trolley busway. They are available in a wide range of sizes and types for lighting loads and such power loads as cranes, hoists, portable tools and high-cycle tools.

A common application of trolley busway is for lighting branch circuits in industrial and some commercial occupancies. In such applications, the trolley busway is generally supported by a hanger or messenger cable and the lighting fixtures are suspended from the busway.

The lighting fixtures are connected to the busway by plug and cord connection. The busway is supplied from a branch circuit protective device in a lighting panelboard.

Current Limiting Busway

Still another type of busway is **high-reactance busway** which is used to limit short circuit currents in modern high capacity electrical systems. This busway is made in rating from 1,000-amps to 4,000-amps, for 3-phase systems rated 600 volts or below.

The busway can be used as service entrance conductors where it can reduce the available short circuit current to a value within the rating, interrupting capacity, of system protective devices. It may also be used for feeders to limit short circuit currents

Connectors
and
Terminals

CONNECTING and terminating devices for electrical wires and cables are the devices used to connect conductors together and to the other components which make up the distribution system. A well designed electrical distribution system made up of the highest grade of equipment can actually be made ineffective by poor practice in the selection and application of connectors and terminals.

The NE Code

Connectors are used in electrical circuits to provide tight, effective bonding between conducting parts and to assure a continuous low-resistance path for current flow in accordance with the given circuit capacities. This important function is covered in the National Electric Code—"Connection of conductors to terminal parts shall insure a thoroughly good connection without damaging the conductors and shall be made by means of pressure connectors (including set screw type), solder lugs, or splices to flexible leads except that No. 8 or smaller solid conductors and No. 10 or smaller stranded conductors may be connected

by means of clamps or screws with terminal plates having upturned lugs."

The code also covers electrical splices: "Conductors shall be spliced or joined with approved splicing devices or by brazing, welding or soldering with a fusible metal or alloy. Soldered splices shall first be so spliced or joined as to be mechanically and electrically secure without solder and then soldered. All splices and joints and the free ends of conductors shall be covered with an insulation equivalent to that on the conductors."

Connectors are made in a very wide range of types and sizes depending upon the size of conductor, the voltage of the circuit, the specific connection to be made and the method by which the conductor is attached to the connector. There are solderless types, which have become very popular through recent years, and the solder type.

Small Wires

Probably the most common connectors are those used for branch circuit wiring and low-capacity feeders. These are the small insulated splicing connectors used to connect

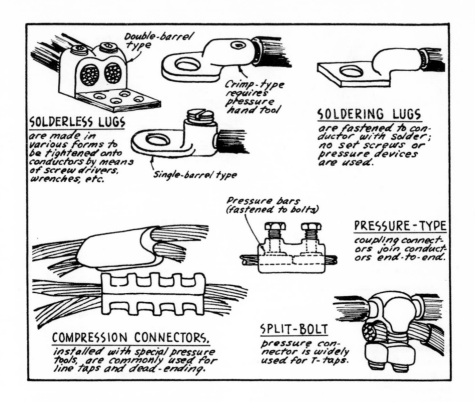

SOLDERLESS LUGS *are made in various forms to be tightened onto conductors by means of screw drivers, wrenches, etc.*

Double-barrel type

Single-barrel type

Crimp-type requires pressure hand tool

SOLDERING LUGS *are fastened to conductor with solder; no set screws or pressure devices are used.*

Pressure bars (fastened to bolts)

PRESSURE-TYPE *coupling connectors join conductors end-to-end.*

COMPRESSION CONNECTORS, *installed with special pressure tools, are commonly used for line taps and dead-ending.*

SPLIT-BOLT *pressure connector is widely used for T-taps.*

circuit wires to lighting fixture wires in outlets, to splice and tap wires in switch and receptacle boxes and to subdivide a circuit home run into a number of legs. Typical connectors of this type are used for wires ranging in size from #18 to #10 on circuits up to 600 volts.

A number of different mechanical connection methods are used on these connectors, including the following: a. **Screw-on type,** which consists of an insulated (plastic or bakelite) cap with an internal threaded core and coil spring to squeeze inserted wire ends into a tight joint by simply screwing on the splice; b. **Set screw type,** which consists of a small tubular element with a set screw to grip the wire ends inserted into it and an insulated cap to screw over the set screw element; c. **Pressure type,** which consists of a small tubular copper cap for covering the bare wire ends and holding them

under pressure applied by special crimping pliers, with an insulating cap for use over the connection.

Large Cables

Connectors used for splicing larger wire and cable sizes include cast and formed assemblies made of pure copper or copper alloys. Typical straight-through connectors for splicing cable sizes from #14 to 2000 MCM are tubular type assemblies which take a bare conductor at each end and have internal gripping plates which hold the conductors under pressure exerted by a bolt or screw through the assembly.

There are also connectors for splicing conductors in a parallel position, for tee tap splices (in which the tap conductor is positioned at right angles to the tapped conductor), and for cross splices. These are rugged compression types using one or more bolts

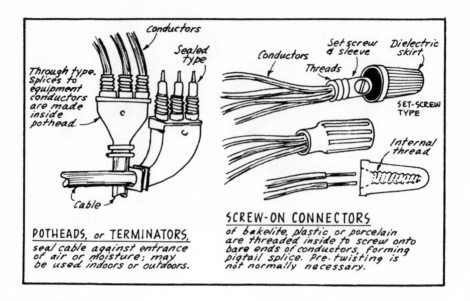

Through type. Splices to equipment conductors are made inside pothead.

Conductors

Sealed type

Cable

POTHEADS, or TERMINATORS, seal cable against entrance of air or moisture; may be used indoors or outdoors.

Conductors

Set screw & sleeve

Dielectric skirt

Threads

SET-SCREW TYPE

Internal thread

SCREW-ON CONNECTORS of bakelite, plastic or porcelain are threaded inside to screw onto bare ends of conductors, forming pigtail splice. Pre-twisting is not normally necessary.

to provide the pressure. Typical applications for these include: feeder taps to panelboards or switches, gutter taps to subfeeders, taps from feeders in wiring troughs, and for service entrance tap connections. In all of these cases, the tap splice assembly is insulated by careful wrapping with insulating tapes.

Because of the particular characteristics of aluminum, connectors for aluminum cable are designed to provide tight electrical bonding for low resistance and mechanical solidity.

Terminal Lugs

Lugs are devices which are connected to ends of wires and cables to facilitate connection to equipment terminals. There are solderless types and solder types, although solderless lugs are the most widely used, Of the solderless types, there are some which are crimped to the bare conductor end by means of a special pliers—a crimping hand tool—and others which have a bolt-driven clamp or pressure plate to grip the conductor. On all lugs, the end opposite that which grips the

conductor is a flat tongue with one or more holes to accommodate bolts used to hold the tongues to equipment terminals.

High Voltage

For electrical connections in circuits rated above 600 volts, proper insulation becomes a much more important consideration than providing the actual continuity of the current path. Splices and terminals must be made carefully in full accordance with instructions of connection specialists.

For high voltage splices, there are a number of splicing kits available based on specific circuit voltage and exact type of cable; i.e., interlocked armor, neoprene-jacketed and lead-covered— for particular conductor insulations.

A common high voltage connecting device is the "pothead." Potheads are used to connect cables to equipment or to other circuits. A pothead is a sealed terminal which provides connection to the conductor in the cable but provides moisture-proofing for the conductor's insulation and seals in cable impregnating oil.

Service Equipment

"**S**ERVICE EQUIPMENT" is a term most commonly used to describe special panelboard assemblies which incorporate service entrance disconnect means and overcurrent protection for the service and for feeder or branch circuits.

Here the term is used to cover the assortment of equipment and hardware used for single-phase services to residential and small commercial occupancies.

This includes: service masts, weatherheads, meter sockets, service switches and service entrance panelboards.

Service Panelboards

These panelboards are designed to simplify service entrance installations up to 200 amps at 120/240 volts, 3-wire, single-phase ac. For a very wide range of applications, such panels provide control and protection for the incoming service feeder and for the circuits into which the incoming power is broken down.

The internal construction is arranged to subdivide the total power capacity of the panel among load circuits. A variety of arrangements is available to meet common circuiting requirements on small services.

One of the most important advantages of service panelboards is that they provide ready compliance with a number of requirements of the National Electrical Code.

Article 230 of the Code requires that a set of service entrance conductors be provided with a "readily accessible means of disconnecting."

For this disconnect, typical service panelboards include a main switch or circuit breaker through which all power must pass to the load circuits.

Six Subdivisions

The NE Code, however, permits the use of up to six switches or six circuit breakers connected in parallel to the service entrance conductors and serving as the disconnect means.

The six devices must be manually operable and located at a readily accessible point nearest to the entrance of the conductors, either inside or outside the building. The six switches or CB's may be mounted in a single enclosure, such as a service entrance panelboard, or each may be mounted in a separate enclosure.

According to the six-subdivision rule of the Code, the disconnecting requirement is satisfied if the service entrance conductors can be completely disconnected from all load circuits by operation of no more than six devices. The Code itself says that complete disconnecting must be made by no more than six operations of the hand.

In typical service entrance panelboards, each of the six devices used to disconnect the service conductors may be a one-pole or two-pole switch or CB. And two single-pole devices (or even three single-pole devices) may be considered to be a single device provided they are equipped with a common tie handle which makes it possible to operate all of the poles with one hand motion.

SERVICE ENTRANCE EQUIPMENT

provides the means for bringing electricity from the utility's lines into the customer's building.

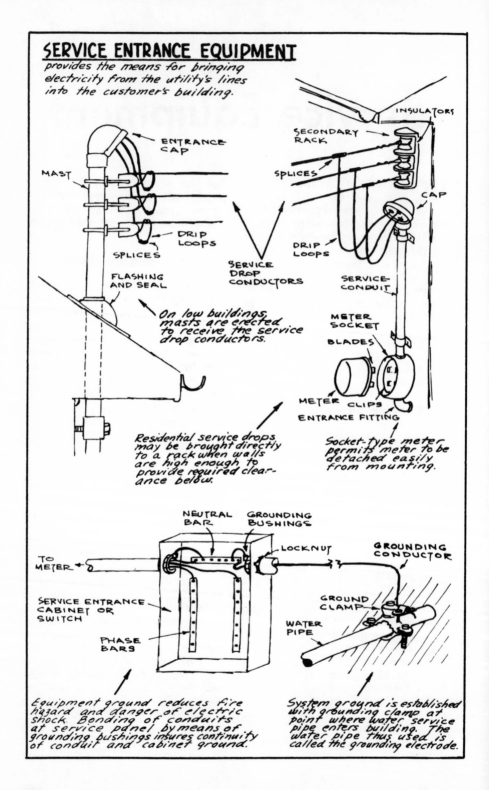

ENTRANCE CAP

MAST

DRIP LOOPS

SPLICES

FLASHING AND SEAL

INSULATORS

SECONDARY RACK

SPLICES

CAP

DRIP LOOPS

SERVICE DROP CONDUCTORS

SERVICE CONDUIT

METER SOCKET

BLADES

METER CLIPS

ENTRANCE FITTING

On low buildings, masts are erected to receive the service drop conductors.

Residential service drops may be brought directly to a rack when walls are high enough to provide required clearance below.

Socket-type meter permits meter to be detached easily from mounting.

NEUTRAL BAR

GROUNDING BUSHINGS

LOCKNUT

GROUNDING CONDUCTOR

TO METER

SERVICE ENTRANCE CABINET OR SWITCH

PHASE BARS

GROUND CLAMP

WATER PIPE

Equipment ground reduces fire hazard and danger of electric shock. Bonding of conduits at service panel by means of grounding bushings insures continuity of conduit and cabinet ground.

System ground is established with grounding clamp at point where water service pipe enters building. The water pipe thus used is called the grounding electrode.

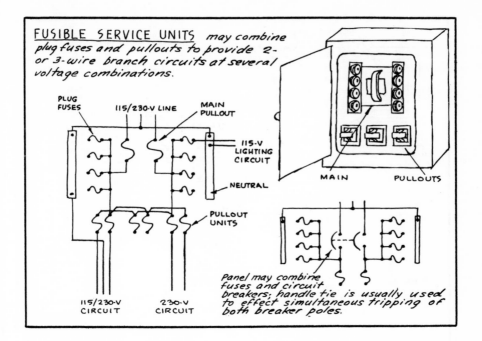

FUSIBLE SERVICE UNITS may combine plug fuses and pullouts to provide 2- or 3-wire branch circuits at several voltage combinations.

PLUG FUSES

115/230-V LINE

MAIN PULLOUT

115-V LIGHTING CIRCUIT

NEUTRAL

PULLOUT UNITS

MAIN

PULLOUTS

115/230-V CIRCUIT

230-V CIRCUIT

Panel may combine fuses and circuit breakers; handle tie is usually used to effect simultaneous tripping of both breaker poles.

Overcurrent Protection

Each ungrounded service entrance conductor must be provided with protection against operating overloads and fault currents due to shorts and grounds. The general rule is that the rating or setting of the protective device in series with each conductor be not greater than the current carrying capacity of the conductor.

One set of fuses or one circuit breaker may be used as the service overcurrent protection where a main switch or CB provides the disconnecting means. However, where up to six switches are used to subdivide the total service capacity, the service overcurrent protection requirement may be satisfied by the total capacity of protective devices (fuses) associated with the switches which supply the load circuits.

Where up to six CB's are used for disconnecting means, the same CB's constitute the service overcurrent protection.

Although the NE Code permits the use of six subdivisions of the service, as described above, this arrangement is disallowed in some sections of the country where a single main disconnect switch is required.

Service panelboards are available in both circuit breaker types and switch and fuse types. Almost any arrangement of appliance circuits and lighting circuits can be quickly assembled from inexpensive, mass-produced components in standardized enclosures.

In typical CB panels, up to five 2-pole CB's may be provided for specific circuits—range, water heater, dryer, clothes washer, etc.—with a sixth 2-pole CB feeding a section of bus from which single-pole CB's tap general purpose lighting and appliance circuits.

In a fuse type panel up to six pullout-type 2-pole fused switches might be used for disconnect and protection, with five of the switches supplying individual appliance circuits

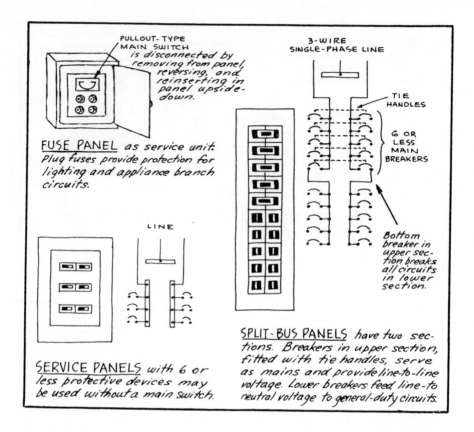

PULLOUT-TYPE MAIN SWITCH *is disconnected by removing from panel, reversing, and reinserting in panel upside-down.*

FUSE PANEL *as service unit. Plug fuses provide protection for lighting and appliance branch circuits.*

3-WIRE SINGLE-PHASE LINE

TIE HANDLES

6 OR LESS MAIN BREAKERS

Bottom breaker in upper section breaks all circuits in lower section.

LINE

SERVICE PANELS *with 6 or less protective devices may be used without a main switch.*

SPLIT-BUS PANELS *have two sections. Breakers in upper section, fitted with tie handles, serve as mains and provide line-to-line voltage. Lower breakers feed line-to-neutral voltage to general-duty circuits.*

and the sixth supplying plug fuse general purpose circuits.

Another advantage of the service panelboard is the engineering integration of the components. Current ratings, interrupting ratings and other factors of compatability of parts are accounted for in the design of the units. This contributes to economy of equipment cost and installation cost.

In any residential wiring layout, the size and makeup of a service entrance panelboard will depend upon the size and layout of the house. Where location of the service entrance is close to the heavy appliance loads—such as laundry and kitchen, and where large capacity home runs to the service can be kept short, a single panelboard can be used for all of the functions of service and distribution. In large homes where the use of a single panelboard at the service location would re-

quire very long home runs for both appliance circuits and general purpose circuits, it is frequently more economical to provide one or more additional panelboards at load centers. Such panelboards can be fed from the service panel by a single feeder, thereby minimizing branch circuit home runs.

A WIDE range of devices and hardware is available for use in residential and other relatively small capacity service entrance arrangements. This includes the equipment used to support and protect the service conductors from the point at which they attach to the building into the point at which the service panel is located, and the equipment needed for effective grounding of the system neutral and the enclosures of equipment. For any overhead residential service from a utility pole line, the details of the service assembly will vary with the type of house and the type of wiring used for the service entrance conductors.

Definitions

In residential overhead services, the conductors from the utility line to the main service switch or panelboard are known as the service conductors. The run of conductors from the connections on the pole to the outside connections on the house are further defined as service drop conductors. The conductors from the outside connections down into the service switch or panel are referred to as the service entrance conductors. The service drop is generally an assembly of two insulated wires with bare neutral. The service entrance conductors may consist of a 3-wire cable or three insulated wires in conduit.

Typical Service

For conventional 2-story houses and split-level houses, the service consists of a service drop to a bracket on the outside of the house, up near the roof. From there, a cable or conduit-and-wire provides the SE (service entrance) conductors down to a meter mounted either outside or inside the house. The meter is a watthour meter which measures the amount of energy consumed. Both indoor and outdoor type meters are made for 2-wire, 120-volt, single-phase services and 3-wire, 240-volt, single-phase services. These meters may be mounted in either a meter socket or a meter trough, which is wired in the SE conductors and provides for ready plugging in of the meter. Meter troughs may be ganged for mounting a number of meters side-by-side.

The conventional method of terminating the SE drop is by means of a service rack which consists of spaced insulators on a metal bracket fixed to the side of the house. The strain of the drop wires is taken by the bracket.

Service Head

When the service entrance wires are contained in conduit, a special fitting is placed on the conduit at the top of its length, adjacent to the service rack. This fitting is called a service head or weatherhead and provides for entry of the SE wires into the conduit while closing the top of the conduit to rain or other water or condensation which might fall on top of the conduit. A wide variety of shapes of service heads is available, but they are all basically the same in that they cover the conduit from the top and allow the wires to come into the head from underneath. Service heads are made with threaded and non-threaded hubs for ready use on thinwall and rigid conduit.

When the SE wires are contained in a SE cable, suitable service heads made especially for cable are used. These have provisions for attaching to the wall of the house and contain clamps to hold the cable in the head.

Hardware

Where SE conductors pass through the wall of a house, from outside to inside, provision is made for this sharp right angle turn whether the wires are in cable or conduit. In the case of cable, metal sill plates attach to the outside of the wall and effectively enclose the bend in the cable, assuring protection at this weak point. In the case of conduit entering the house, weatherproof elbows with removable covers provide the 90° angle in the conduit which is mounted up against the wall.

For installations using SE cable, special clamps are available for attaching the cable to the outside wall of the house. Rubber bushed connectors are also made for SE cable connections to boxes.

A number of devices are made to meet the requirements for grounding at a service entrance. Pressure type ground clamps are made to provide solid connection of grounding conductors of various sizes to water pipes or ground rods. These are bolt type devices which tightly grip the pipe or rod and have connection for the grounding wire. They afford ready compliance with code requirements for grounding of the system neutral on 120- and 240-volt circuits and all metallic non-current-carrying parts of equipment and enclosures.

And because the code requires that all metal parts be bonded together and connected to ground, effective use can be made of grounding bushings and rings to bond together service raceway, meter enclosure, service panel enclosure and other enclosures at the service location, and to connect these parts to the grounding conductor.

Service Masts

With the exception of the service rack arrangement, all of the foregoing details apply to low or ranch type houses as well as to 2-story houses. But for bungalows and ranch houses, termination of the service drop and arrangement of service head frequently require special attention. The National Electrical Code and other building codes and utility codes require a minimum clearance between ground and service drop conductors. The NEC states that the drop must be not less than 10 feet from the ground or from any platform or projection from which it might be reached. And, generally, the point of attachment to the building must not be less than 10 feet.

On ranch houses and other houses with low roofs, a conventional service rack assembly is ruled out because the edge of the roof is lower than 10 feet above the ground. To meet such conditions, special above-roof mast assemblies are used. These are commonly made up in kits of all required parts and hardware to bring the drop into a service head on top of a length of conduit which rises above the roof and extends down through the eave of the roof and along the side of the house. Accessories include: roof flashing and seals, mounting brackets, clamp insulators and any special fittings.

Switchboards
and
Panelboards

INSIDE and outside electrical systems for light and power make use of switchboards and panelboards to the same extent as they make use of wire and cable.

In fact, switchboards and panelboards provide points in electrical systems at which blocks of electric power are broken down from large conductors and apportioned, as required by loads, among smaller conductors.

In addition, they afford a logical place for using switches or circuit breakers as control devices for the flow of energy or for disconnecting loads for safe maintenance.

And still another necessary function which can readily and effectively be accomplished in switchboards and panelboards is that of protection of conductors and equipment against excessive current flow due to overloads, grounds and shorts.

Although switchboards and panelboards both accomplish these results, they differ in construction and application. Switchboards are invariably floor standing assemblies; panelboards are wall mounting.

Switchboards provide breakdown of relatively large blocks of power supplied by service conductors in most cases or by large feeders in extensive, high capacity systems.

Panelboards provide breakdown of power supplied by small service conductors or by feeders from switchboards. Circuits originating in switchboards are generally feeders.

Circuits originating in panelboards may be feeders, subfeeders or—which is the most common case—branch circuits for lighting, convenience receptacles, appliances and/or motor loads.

Switchboards

According to the National Elec-

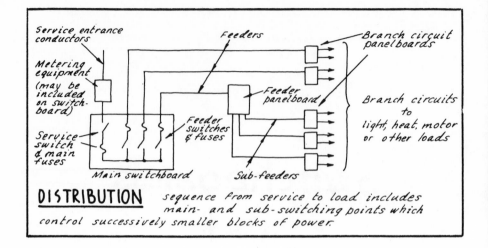

Service entrance conductors

Feeders

Branch circuit panelboards

Metering equipment (may be included on switchboard)

Service switch & main fuses

Feeder switches & fuses

Feeder panelboard

Branch circuits to light, heat, motor or other loads

Main switchboard

Sub-feeders

DISTRIBUTION sequence from service to load includes main- and sub-switching points which control successively smaller blocks of power.

trical Code section on definitions, a switchboard is a large single panel, frame, or assembly of panels, on which are mounted—on the face or back or both—switches, overcurrent and other protective devices, buses, and, usually, instruments.

This definition is well suited to the types of switchboards in common use today for systems rated 600 volts and below—commonly referred to as low voltage systems.

Common usage, however, reserves the word "switchgear" for switchboards used on voltages above 600 volts—2300 volts and above. This derives from the more rugged construction and the commonly mechanized internal assembly.

Switchboards in common use today may be divided into several types according to construction. There are open types (live-front and dead-front types) and enclosed, safety types of boards.

Open type switchboards consist of switching and protective devices mounted on a large vertical panel which is readily accessible from both front and back. The protective devices may be fused switches or circuit breakers.

The rear of such boards is commonly enclosed with a high grillwork fence to minimize the hazard due to

the energized equipment. Access to the rear of such switchboards is provided by a locked door in the grill enclosure. Only qualified personnel have keys to this door.

Open type boards are limited in modern design practice to systems operating at 600 volts and below. In this range, however, some design engineers actually prefer open type boards to the completely enclosed, safety type.

They feel that open boards have better ventilation, offer easier and more effective maintenance and can be safely applied. Nevertheless, the predominant trend over recent years has been toward use of enclosed types of switchboards.

A further breakdown of open type boards can be made according to manner in which the switching and protective devices are mounted on the face of the board.

If this equipment is mounted so a person standing in front of the board can make contact with any current carrying parts of the equipment, the switchboard is a live front type. These boards are generally arranged with knife switches and fuses on their front panel. Such boards present the danger of contact with these hot parts.

For this reason, the National Electrical Code requires that switchboards which have any exposed live parts be located in permanently dry locations and only where they are under competent supervision and accessible only to qualified personnel.

If all devices mounted on the board have no current-carrying parts accessible from the front, the board is known as a dead front type. In such boards, only the operating handles are accessible from the front.

The most common type of switchboard used in modern electrical systems for commercial, industrial and institutional buildings is the completely enclosed, safety switchboard. Such boards consist of fused switches or circuit breakers mounted as unit assemblies in metal, floor standing cabinet enclosures.

The front surface of these boards is completely free of live parts, with only operating handles or pushbuttons accessible for operation of the enclosed devices.

In many installations, switchboards are equipped with meters on their front panels to advise operating personnel of various conditions of load. These may include: voltmeters, ammeters, wattmeters, kilowatthour meters, power factor meters and var meters.

Switchboards also commonly contain current transformers for utility meterings of electrical energy consumption, for metering by the owner and for various control and protecttion purposes. Still other boards may contain automatic transfer switches which disconnect the board from the normal utility supply when that supply fails and connects it to a source of emergency supply, such as an engine generator.

A wide variety of special switchboards are commonly found in modern industrial and commercial practice—such as boards for hi-cycle lighting and for heavy dc loads.

Switchboards used with generator supplies usually include frequency meters and other instruments used for operating one or more generators. In most cases, meters are used only in the main switchboard of a system. Sub-distribution switchboards generally do not have meters.

THE NEC defines a panelboard as a "single panel or a group of panel units designed for assembly in the form of a single panel; including buses, and with or without switches and/or automatic overcur-

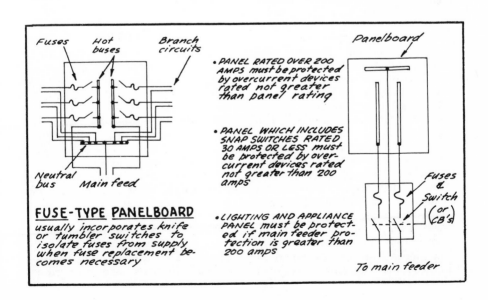

Fuses   Hot   Branch
         buses circuits

Neutral
bus   Main feed

FUSE-TYPE PANELBOARD
usually incorporates knife or tumbler switches to isolate fuses from supply when fuse replacement becomes necessary

• PANEL RATED OVER 200 AMPS must be protected by overcurrent devices rated not greater than panel rating

• PANEL WHICH INCLUDES SNAP SWITCHES RATED 30 AMPS OR LESS must be protected by overcurrent devices rated not greater than 200 amps

• LIGHTING AND APPLIANCE PANEL must be protected if main feeder protection is greater than 200 amps

Panelboard

Fuses & Switch (or CB's)

To main feeder

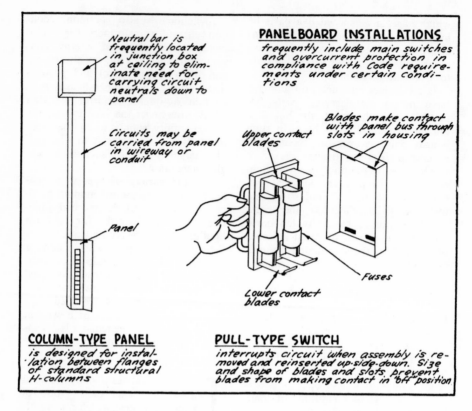

Neutral bar is frequently located in junction box at ceiling to eliminate need for carrying circuit neutrals down to panel

Circuits may be carried from panel in wireway or conduit

Panel

PANELBOARD INSTALLATIONS frequently include main switches and overcurrent protection in compliance with Code requirements under certain conditions

Blades make contact with panel bus through slots in housing

Upper contact blades

Lower contact blades

Fuses

COLUMN-TYPE PANEL
is designed for installation between flanges of standard structural H-columns

PULL-TYPE SWITCH
interrupts circuit when assembly is removed and reinserted up-side-down. Size and shape of blades and slots prevent blades from making contact in "off" position

rent protective devices for the control of light, heat or power circuits of small individual as well as aggregate capacity; designed to be placed in a cabinet or cutout box placed in or against a wall or partition and accessible only from the front."

Switchboards are relatively large, floor-standing cabinet assemblies; panelboards are smaller cabinet assemblies designed for surface or recessed mounting on walls or partitions. Switchboards are installed in special electrical equipment rooms and are accessible only to authorized personnel. Panelboards are installed throughout buildings, in halls, offices, manufacturing areas, etc.—and are commonly accessible to all building occupants for use of the switches or circuit breakers to control lights or other branch circuit loads.

Panelboards are used at points of electric power subdivision where the

capacity of the supply does not exceed 600 volts. Panelboards are limited to use in systems rated 600 volts and below.

Types

A wide range of panelboards is made for single-phase and three-phase circuits, for use at small service entrances for subdividing the incoming supply, for use in feeders to subdivide into subfeeders and for use in feeders or subfeeders to subdivide into branch circuits for lighting loads, motor loads and other power loads. Residences, stores, small schools and similar occupancies with small capacity services use panelboards for service equipment.

The term "power panelboards" refers to panelboards which have relatively high capacity main buses— 200,400,600 amps—and are used to break a feeder down into subfeeders or to break a feeder or subfeeder down into branch circuits for heavy motor loads in commercial and industrial buildings.

Power panels are commonly used for handling widespread concentrations of power loads on 240-and 480 volts delta systems, although they are also used on wye connected systems.

Lighting and appliance panelboards, with ratings of main buses at 200 amps and below, are the most widely and commonly used type of panelboard. These are the panels at which branch circuits are derived from feeders or subfeeders to supply general area fluorescent and/or incandescent lighting equipment, small appliance loads and convenience outlets for

cord connected devices and appliances.

Panelboards—both power panels and lighting and appliance branch circuit panelboards—are available with either circuit breakers or fuses for overcurrent protection. Either type of protective device affords the required protections for the smaller conductors derived from the bus bars in the panel. These bars are supplied by a set of feeder conductors having current-carrying capacity at least equal to the loads to be served. In fuse type panels, some type of switching device is generally provided for each set of smaller conductors to permit killing the circuit before changing a fuse. There are fuse panels, however, which do not include switch-

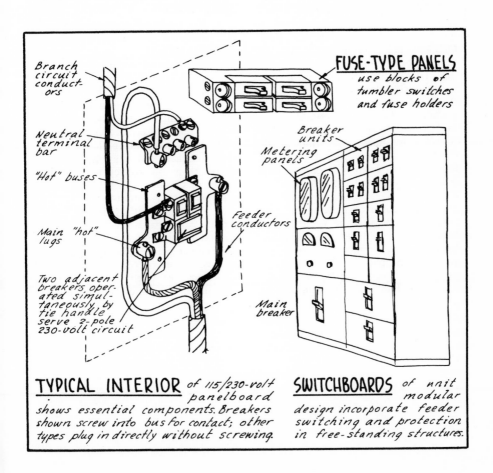

Branch circuit conductors

Neutral terminal bar

"Hot" buses

Main "hot" lugs

Two adjacent breakers operated simultaneously by tie handle serve 2-pole 230-volt circuit

FUSE-TYPE PANELS use blocks of tumbler switches and fuse holders

Breaker units

Metering panels

Feeder conductors

Main breaker

TYPICAL INTERIOR of 115/230-volt panelboard shows essential components. Breakers shown screw into bus for contact; other types plug in directly without screwing.

SWITCHBOARDS of unit modular design incorporate feeder switching and protection in free-standing structures.

ing devices. Such panels must be inaccessible to unauthorized personnel. Some panels use tumbler type switches, others use knife type switches, and still others use pull-out type disconnecting devices.

Circuit breaker panelboards provide overcurrent and short circuit protection for the circuits as well as on-off control of each of the circuits.

Special Assemblies

In modern electrical systems, a number of special types of panelboard arrangements are commonly used. These include panelboards which have magnetic contactors in their enclosure to supply the complete bus or a section of the bus in the enclosure. A typical application would include a lighting panelboard with a remote control switch connected to the bus. The circuits in the panel might supply parking lot lighting or other lighting units outside the building. The control circuit of the contactor (or remote control switch) is then carried from the panel to a pushbutton located in a guard's office or adjacent to an outside door. This pushbutton affords operation of the contactor, and through it the lighting circuits, from a point remote from the panel but close to the area to be lighted.

Another form of panelboard using contactor control is the **split-bus panelboard.** In this assembly, the feeder to the panel connects to bus bars which extend only part of the way up in the panelboard. The top end of this section of bus feeds through a mag-

netic contactor to a second section of the bus which extends to the top of the panel enclosure. Circuits connected to the lower section of bus are controlled either by their branch circuit switching devices or by local control out on the circuit. The magnetic contactor exercises on-off control for the top section of bus, thereby controlling the circuits connected to that section of the bus.

The Code

The NE Code specifically covers installation and application of panelboards. In many systems, these requirements of the code are very important to proper application of panelboards. Typical code regulations are as follows:

1. To facilitate comprehension of the rules, the code sets forth a very technical definition of a lighting and appliance branch circuit panelboard. This is the general purpose type panel and is defined as "one having more than 10% of its overcurrent devices rated 30 amperes or less, for which neutral connections are provided."

2. A lighting and appliance branch circuit panelboard may not contain more than 42 overcurrent devices (fuses or circuit breaker poles) in any one cabinet or cutout box.

3. A lighting and appliance branch circuit panelboard which is supplied by conductors protected at more than 200 amperes must be equipped with its own main overcurrent protection (fuses or CB) rated for the carrying capacity of the panel's bus bars.

Explosion-Proof Equipment

ELECTRICAL DESIGN and construction for hazardous locations requires careful selection and installation of equipment and materials. All enclosures—motor frames, luminaire housings, cabinets for panelboards and switching devices, receptacle outlets, conductor enclosures —must be suitable for use in the particular type of hazardous location.

Strict conformity to the provisions of the National Electrical Code and reference to the "Hazardous Location Equipment" list of the Underwriters' Laboratories, Inc., can effectively minimize the hazards of such application.

The principal objective of all hazardous equipment design is to prevent any heat, arcs and sparks from electrical devices from igniting the highly combustible gases and particles which are normally present in areas classified as hazardous.

Classifications

Articles 500 to 517 of the National Electrical Code cover general and specific requirements on wiring in hazardous locations. The basic idea behind these requirements is that parts of electrical systems in or passing through areas where flammable gases, combustible dusts or ignitible fibers or flyings are or may be present must be of such construction that arcs or sparks cannot be transmitted out of the equipment or conductor enclosures to cause ignition of any hazardous atmospheric mixture. In setting up its rules, the Code divides hazardous locations into three classes which cover the range of hazardous atmospheres, as follows:

CLASS I LOCATIONS — those in which flammable gases or vapors are or may be present in the air in quantities sufficient to produce explosive or ignitible mixtures. Such locations are further subdivided according to the characteristics of the type of gas or vapor which produces the hazardous atmospheric mixture. This breakdown is made for the purpose of testing and approving equipment for application in specific atmospheres. Underwriters' Laboratories tests and approves Class I equipment for atmospheres—**Group A,** acetylene; **Group B,** hydrogen; **Group C,** ethyl-ether, ethylene or cyclo-propane gases;

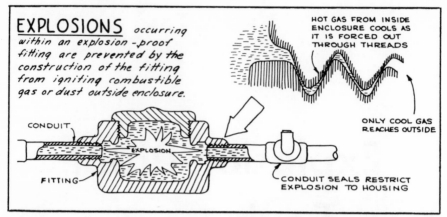

EXPLOSIONS occurring within an explosion-proof fitting are prevented by the construction of the fitting from igniting combustible gas or dust outside enclosure.

HOT GAS FROM INSIDE ENCLOSURE COOLS AS IT IS FORCED OUT THROUGH THREADS

ONLY COOL GAS REACHES OUTSIDE

CONDUIT

FITTING

EXPLOSION

CONDUIT SEALS RESTRICT EXPLOSION TO HOUSING

Group D, gasoline, naphtha, benzine, butane, propane, alcohol, acetone, lacquer solvent vapors or natural gas.

When a hazardous location has been determined as Class I and the group designation has been determined by the particular atmosphere, the location must then be identified as either Division 1 or Division 2. This further distinction indicates the degree of the hazard. Division 1 locations have hazardous concentrations of the gas or vapor continuously, intermittently or periodically under normal conditions. Division 2 locations are those in which the volatile liquids or gases are handled in closed containers or closed systems and are, therefore, less hazardous locations than Division 1.

CLASS II LOCATIONS — those where the presence of combustible dusts present a fire or explosion hazard. Based on the type of dust, equipment for such locations are further designated—Group E, metal dusts; Group F, carbon black, coal or coke dust; Group G, flour, starch or grain dust.

Class II, Division 1 locations are those where dust is in suspension in the air continuously, intermittently or periodically — in quantities sufficient for fire or explosion. Class II, Division 2 locations are those where the dust is not in suspension but where it may collect on electrical equipment and produce a fire hazard.

CLASS III LOCATIONS — those where easily ignitible fibers or flyings are present but not likely to be suspended in the air in quantities sufficient to produce ignitible mixtures. In this class, Division 1 locations are those where ignitible fibers or materials producing combustible flyings are handled, manufactured or used. Division 2 locations are those where such fibers are stored or handled. There are no group designations for Class III locations.

Application

Equipment used in Class I and Class II locations should be specifically designed and approved for both Class and Group of hazard. Such equipment should bear the Underwriters' Laboratories label, noting specific approval. This applies to lighting fixtures, motors, controllers, switches, circuit breakers, plugs, receptacles, panelboards and other equipment.

For wiring in hazardous locations, rigid conduit of steel or aluminum may be used with threaded fittings as the only Code-approved raceway system in all hazardous areas. Such conduit provides required mechanical protection and contains internal explosions to prevent propagation of the flame to the hazardous atmosphere external to it. Fittings and junction boxes used with conduit systems must be "explosion-proof" design in Class I locations; in Class II locations, they must be "dust-tight" or "dust-ignition-proof" types.

The NE Code covers specific details on wiring methods, use of sealing fittings, construction of equipment and grounding—for each Class and Division of hazardous location. Sealing fittings are used in conduits in hazardous locations to prevent passage of gases, vapors or flames from one part of the electrical system to another part through the conduit. Such fittings provide for the pouring of a compound which fills the cross section of the conduit around the enclosed conductors, effectively sealing off the conduit at that point.

Generally, seals are installed in conduits entering enclosures for switches or other devices which may produce arcs, sparks or high temperatures. And the seal in each conduit must be placed within 18 inches of the point of entrance of the conduit to the enclosure. A seal must also be installed in a conduit run where it passes from a Class I or Class II hazardous location into a non-hazardous area. To facilitate compliance with sealing requirements, some explosion-proof control devices and panelboards are made with integral seals, eliminating need for conduit seals.

Hazardous location luminaires are constructed to prevent entry of gas, vapor or duct into the lamp assembly. And their temperature of operation must not reach the ignition level for the particular atmosphere. Luminaires are labeled for their specific approved application. Some such units also incorporate integral seals.

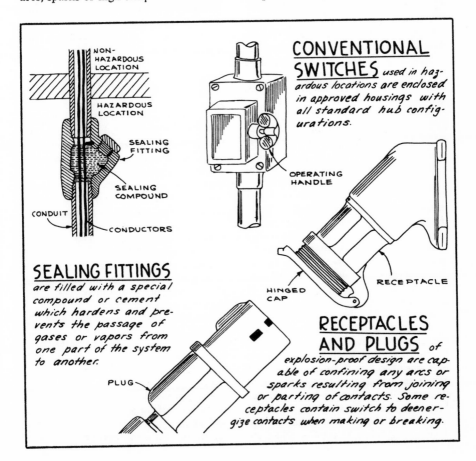

NON-HAZARDOUS LOCATION

HAZARDOUS LOCATION

SEALING FITTING

SEALING COMPOUND

CONDUIT

CONDUCTORS

CONVENTIONAL SWITCHES used in hazardous locations are enclosed in approved housings with all standard hub configurations.

OPERATING HANDLE

RECEPTACLE

HINGED CAP

SEALING FITTINGS are filled with a special compound or cement which hardens and prevents the passage of gases or vapors from one part of the system to another.

PLUG

RECEPTACLES AND PLUGS of explosion-proof design are capable of confining any arcs or sparks resulting from joining or parting of contacts. Some receptacles contain switch to deenergize contacts when making or breaking.

Rectifiers

A RECTIFIER is a device which converts alternating-current power into direct-current power.

Today, over 90 per cent (and the percentage is still growing) of electrical energy is generated and transmitted as alternating-current, offering advantages which dc can never match. Although most of this energy is consumed by ac utilization devices—ac motors, appliances and lighting—there are still many applications in which dc power is essential or distinctly superior to ac power. Such applications are: electroplating, separation of metals, battery-charging, street railways, elevator motors and drives for printing presses. Rectifiers make possible the supply of dc power for these applications.

Although there are many differences among types of rectifiers, including the actual operating principle, they all produce the same effect, i.e., rectification of alternating current.

Rectification is the action which takes place in a rectifier. It refers to the fundamental characteristic of the device by which current is allowed to flow in only one direction through the rectifier. When an ac voltage is applied to a rectifier, current will flow only during that half of the ac cycle when the ac voltage is in the direction of current passage. During the other half cycle, when the voltage creates electrical pressure in the opposite direction, no current will flow through the rectifier.

• **Mechanical Rectifiers**—There are several types of mechanical rectifiers. All accomplish conversion from ac to dc in the same general way: by reversing connection between the dc output bus bars and the incoming ac supply wires in time with the reversals of the ac voltage. The idea is to have the dc positive output bus connected to an ac incoming line only during those halves of the ac cycle when the current will flow from the ac line to the positive bus. When the current reverses in the ac lines, the connections are reversed to keep current flow from ac line to positive bus and from negative bus to ac line. This changing of connections in time with reversal of the ac cycle to maintain direct current flow may be done with a special type of commutator, with cam-actuated contacts or by vibrator action.

Mechanical rectifiers have found industrial and electro-chemical applications, providing as high as 10,000 amperes at voltages between 50 and 600 volts, with efficiencies of over 95 per cent.

• **Copper-Oxide Rectifiers**—A layer of cuprous oxide on a thin piece of copper will allow current to flow through the copper in only one direction.

RECTIFICATION

AC Supply

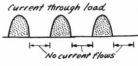

Current through load

ac dc Load Rectifier

No current flows

An ac supply reverses direction every half-cycle, as shown above. If it is connected across a circuit containing a rectifier and a resistance load, current will flow only half the time.

The rectifier permits the current to flow in one direction only; the negative half of the wave is cut off. This result is called "half-wave" rectification. Direct current flows through the load.

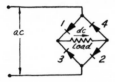

ac

1 dc 4
Load
3 2

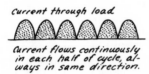

Current through load

Current flows continuously in each half of cycle, always in same direction.

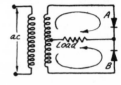

ac

A
Load
B

Four rectifiers may be arranged in a bridge circuit to permit current to flow during each half-cycle, producing "full-wave" rectification. During the (+) half of the ac cycle, current flow is through rectifier 1, load, and rectifier 2. During the (-) half of the cycle, current flow is through rectifier 3, load, and rectifier 4. Load current is always in same direction.

Full-wave rectification may be obtained also by using two rectifiers and a transformer. The load is connected to a center-tap of the transformer secondary. During the (+) half of the ac cycle, current flow is through rectifier A and the load. During the (-) half of the cycle, the flow is through rectifier B and the load, in the same direction.

A typical copper-oxide rectifier consists of an assembly of washers, about 1½ inches in diameter.

Copper-oxide rectifiers are used in instruments, controls, battery charges.

• **Selenium Rectifiers**—A thin layer of selenium will pass current in only one direction when it is placed between and in contact with two metallic electrodes. The basic selinium rectifier unit is called a cell. Cells are connected in various ways to form stacks, giving a wide range of possible voltage and current ratings. Selinium rectifiers are rugged and reliable; they require practically no maintenance, are quiet, and have long life. Efficiency is up around 85 per cent.

Applications of selinium rectifiers are: arc welding, battery charging, burglar alarms, control devices, instruments, electronic equipment, elevator controls, high voltage testing, dc light and power, electroplating.

Typical ratings on selenium rectifiers are as follows: 26-30 volts dc adjustable output from 208/220/440/ or 550 volts ac, 3-phase input, up to 1000 amps, for engine starting or battery charging; 115/230 volts dc from 220/440 volts ac,

3-phase input, 100 kw, for motor loads; 6 volts dc from 220/440 volts ac, 3-phase input, in dc amperes from 250 to 5000, for electroplating.

• **Germanium Rectifiers**—These are relatively new in the field of power rectifiers. Their operating principle is similar to that of the selenium type.

Germanium rectifiers are used to supply power to dc motors, excitation current for synchronous motors, and energy for magnetic chucks, brakes and clutches. The big advantage of the germanium rectifier is its 94 per cent efficiency, even at half load, which represents a new high for static rectifiers. Ratings of typical germanium rectifier assemblies are: 65/125 volts dc, 400/200 amperes dc.

• **Hot-Cathode Rectifiers**—These are tube type rectifiers which contain a heated cathode at one end and a cold anode at the other. Some are high-vacuum tubes, as in the case of radio and TV rectifier tubes, others are gaseous rectifiers. These rectifiers are used principally in battery chargers.

• **Mercury-Arc Rectifiers**—These rectifiers operate in much the same way as the hot cathode rectifiers. They have a mercury pool which serves as the hot cathode and does not deteriorate. Current can flow through them in only one direction.

A glass-tube mercury-arc rectifier has been used for many years in chargers for storage batteries. Multianode type mercury-arc rectifiers are in common use for many power applications.

• Thyratrons—In general, these are similar to hot-cathode gaseous rectifier tubes, with the addition of another element, a control grid. Thyratron rectifiers generally have low power ratings. Typical industrial applications are: regulators for alternator fields, lamp dimming, timing of the current in welding operations, and in other cases requiring rectification, relaying and control.

• Ignitrons—The ignitron is a mercury-arc type rectifier and consists of a case with end connections, a graphite main anode and a mercury pool cathode. Single ignition units may be assembled in groups and connected in 3-phase, 6-phase or 12-phase combinations, just as

the anodes of a single tank are connected. In the smaller ratings, ignitrons are sealed in glass tubes; in intermediate ratings, sealed-in water-cooled metal tanks are used for the ignitrons; in the large power assemblies, pumped all-steel water-cooled tanks are used.

Features of ignitron rectifier assemblies are: high efficiency, quiet operation, low maintenance, no special air cooling required and unattended operation. Typical applications of ignitron assemblies are: dc motor loads, heavy industrial electrolytic work, steel mill drives, mining service, railway service. The sealed tube types have ratings from 75 to 1000 kw, 125 to 600 dc volts. The pumped types go as high as 6000 kw, at 600 dc volts. In the reduction of aluminum, 850 volt dc buses are supplied with over 100,000 amperes, from an ignitron rectifier assembly.

• Excitrons—These are single-tank mercury-arc rectifiers which vary somewhat from the ignitron. Their characteristics and application are about the same.

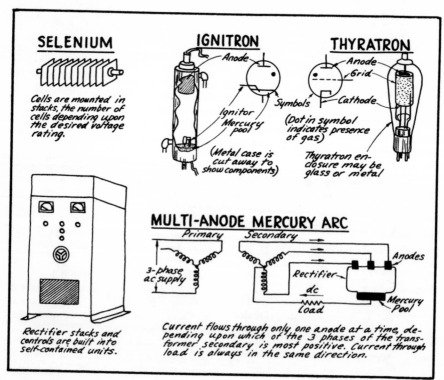

SELENIUM

Cells are mounted in stacks, the number of cells depending upon the desired voltage rating.

Rectifier stacks and controls are built into self-contained units.

IGNITRON

Anode

Ignitor

Mercury pool

Symbols

(Metal case is cut away to show components)

THYRATRON

Anode

Grid

Cathode

(Dot in symbol indicates presence of gas)

Thyratron enclosure may be glass or metal

MULTI-ANODE MERCURY ARC

Primary   Secondary

3-phase ac supply

Rectifier dc

Anodes

Mercury Pool

Load

Current flows through only one anode at a time, depending upon which of the 3 phases of the transformer secondary is most positive. Current through load is always in the same direction.

Converters

IN ADDITION to rectifiers—which convert alternating-current power to direct-current power—there are other devices which can convert electric power from one form to another form. These devices are commonly labeled "converters." There are motor-generator converters, synchronous converters, frequency converters, phase converters and a wide range of small plug-in unit converters for very low power applications.

M-G Sets

Motor-generator sets are capable of converting large amounts of electric power from ac to dc or from dc to ac. Such sets consist of a motor and a generator arranged end-to-end, with their rotor shafts firmly coupled together. To convert from ac to dc, the motor must be an ac type and the generator must be a dc unit. To convert from dc to ac,

the motor must be dc and the generator ac. In both cases, input power is fed to the motor; output power is taken from the generator.

M-g sets used for converting large amounts of ac power to dc are made with either an induction motor or a synchronous motor driving the generator. Each motor type has certain advantages, depending upon the application.

A common application of motor-generator converters is in commercial and industrial emergency power and lighting systems. A typical setup might have a dc motor and an ac generator. In the event of failure of the normal ac supply, an automatic switch will connect storage batteries to the dc motor which in turn will drive the ac generator and supply ac power to the emergency circuits. Another arrangement might have an ac motor driving a dc generator which is charging

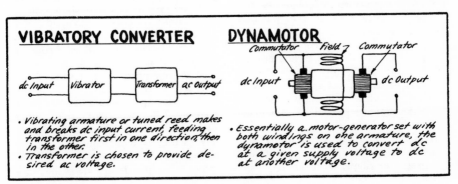

VIBRATORY CONVERTER

dc Input — Vibrator — Transformer — ac Output

- Vibrating armature or tuned reed makes and breaks dc input current, feeding transformer first in one direction, then in the other.
- Transformer is chosen to provide desired ac voltage.

DYNAMOTOR

Commutator   Field   Commutator
dc Input — dc Output

- Essentially a motor-generator set with both windings on one armature, the dynamotor is used to convert dc at a given supply voltage to dc at another voltage.

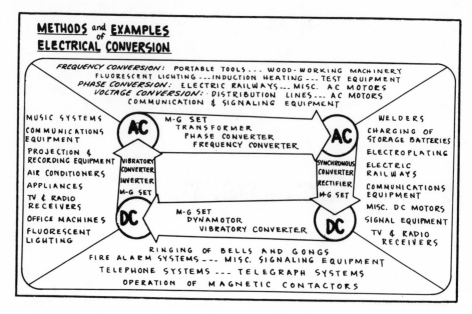

METHODS and EXAMPLES of ELECTRICAL CONVERSION

FREQUENCY CONVERSION: PORTABLE TOOLS... WOOD-WORKING MACHINERY
FLUORESCENT LIGHTING --- INDUCTION HEATING --- TEST EQUIPMENT
PHASE CONVERSION: ELECTRIC RAILWAYS... MISC. AC MOTORS
VOLTAGE CONVERSION: DISTRIBUTION LINES --- AC MOTORS
COMMUNICATION & SIGNALING EQUIPMENT

MUSIC SYSTEMS
COMMUNICATIONS EQUIPMENT
PROJECTION & RECORDING EQUIPMENT
AIR CONDITIONERS
APPLIANCES
TV & RADIO RECEIVERS
OFFICE MACHINES
FLUORESCENT LIGHTING

AC

VIBRATORY CONVERTER
INVERTER
M-G SET

M-G SET
TRANSFORMER
PHASE CONVERTER
FREQUENCY CONVERTER

M-G SET
DYNAMOTOR
VIBRATORY CONVERTER

DC

AC

SYNCHRONOUS CONVERTER
RECTIFIER
M-G SET

WELDERS
CHARGING OF STORAGE BATTERIES
ELECTROPLATING
ELECTRIC RAILWAYS
COMMUNICATIONS EQUIPMENT
MISC. DC MOTORS
SIGNAL EQUIPMENT
TV & RADIO RECEIVERS

DC

RINGING OF BELLS AND GONGS
FIRE ALARM SYSTEMS --- MISC. SIGNALING EQUIPMENT
TELEPHONE SYSTEMS --- TELEGRAPH SYSTEMS
OPERATION OF MAGNETIC CONTACTORS

storage batteries used for emergency lighting.

Synchronous Converter

A disadvantage of the motor-generator set is that it requires two machines, taking up considerable floor space and presenting a twofold maintenance consideration. These problems are eliminated by the synchronous converter, also called the rotary converter. The synchronous converter is a single machine which can convert large amounts of ac power to dc power, and vice versa. Compared to the motor-generator converter, it is more efficient and economical.

The synchronous converter consists of a rotating armature with a commutator, slip rings connected to the armature and a winding structure to provide the necessary magnetic field. In a typical unit, alternating-current is supplied to the slip rings and direct-current is taken from the commutator through carbon brushes. When operated in this way, the machine is said to be a direct synchronous converter. Another unit might have dc power supplied to the commutator through the brushes and ac power taken from the slip rings. This unit is commonly called an inverted synchronous converter or simply an inverter.

Frequency Converters

In modern commercial and industrial electrical systems, it is often desirable to change the frequency of alternating-current power to meet the demands of a special application. There are many cases where electric power can be more effectively, efficiently and economically utilized at a higher or lower frequency than that frequency at which it is available. For instance, induction heating equipment generally requires ac power at a frequency anywhere between 500 and 10,000 cycles, although the plant will have a 60-cycle ac distribution system. Other applications which have to be powered from a 50 or 60-cycle ac system are: high-frequency power tools, 120 and 180 cycles; electric traction (streetcars, trains), 25 cycles; high-cycle motor testing equipment, 360 cycles and higher.

A frequency converter or frequency changer, as it is often called, is a device or combination of devices which will take an ac input at some frequency and provide an ac output at some other frequency. Frequency converters are of two types: rotating type and static type. The rotating type of frequency converter is generally an m-g set in which the motor is designed to operate at the available frequency of ac supply and the generator is designed to provide power output at the desired frequency. In a typical installation, this might mean a 60-cycle motor fed from the distribution system, driving a generator with a 400 cycle ac output which is operating high-frequency fluorescent lighting equipment.

Static frequency converters use mercury-arc rectifying devices in special circuits and have been used for high power conversions from 60 to 25 cycles and for converting from low-frequency ac to energy at several thousand cycles, as used in induction heat treating of metals.

Phase Converters

Another type of converter is the phase converter. This device is used to change single-phase ac to three-phase ac, and sometimes vice versa. It may be either a rotating machine or a static device. And a rotating phase converter may be either an m-g set or a single-machine converter.

An m-g set used for single to three-phase conversion would, of course, consist of a single-phase motor driving a three-phase generator. A single-machine phase converter is actually an induction motor to which single-phase power is fed and from which three-phase power is taken. This is commonly used in electric railway applications.

At least two companies are making static phase converters, using transformers and capacitors. These compact, efficient devices are currently finding widespread application in farm and other areas where only single-phase ac power is available. In such areas, the phase converter makes possible the use of three-phase motors, which are better suited to many jobs than are single-phase motors.

It should be noted here that simple transformer hookups are capable of converting ac power from two-phase to three-phase and from three-phase to six-phase.

Converter "Packages"

In addition to the large power commercial and industrial types of converters described above, there are many small portable plug-in units which are also called converters. These are usually electronic, rectifier or vibrator devices housed in small ventilated steel enclosures and equipped with connecting cables. Such units are used to convert ac to dc, low-voltage dc to high-voltage dc or low-voltage dc (6 to 12 vdc) to 120 vac. All of these are low power units, used for such applications as electroplating, battery charging, radio and electronic power supplies and operation of ac appliances from dc circuits.

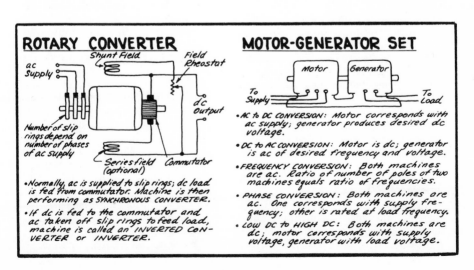

Capacitors

A CAPACITOR is an electrical circuit device which has characteristics suited to a wide range of applications in alternating-current circuits. In electronic circuits, small capacitors are used: to remove the ac ripple from direct-current voltages obtained through the use of rectifiers; to allow passage of ac signals from one circuit to another while preventing the passage of dc; and to provide selection of particular frequencies of signals, as in radio and TV tuners. In single-phase capacitor type motors, capacitors are used to simulate two-phase operation and thereby provide sufficient starting torque (rotating force required to get the motor going). In fluorescent lamp ballasts, they are used to improve power factor, reducing the amount of current which the ballasts draw from the line. In large power applications, larger sizes of capacitors provide power factor correction with all of its consequent advantages.

Operation

Basically, a capacitor consists of two metallic elements, frequently called "plates," which are separated by an insulating material called a "dielec- tric." The two plates and the dielectric between them make up the bulk of the capacitor. The two plates are connected to external terminals, on the outside of the capacitor case, for connecting the capacitor to the circuit in which it will be used.

A capacitor is essentially an electric storage device which can be used to take energy from an ac line during half of the ac cycle and to return energy to the line during the other half of each cycle. In a circuit, a capacitor acts very much like a spring: when voltage forces current into a capacitor during one half of the cycle, the capacitor stores the energy represented as a compressed spring stores energy; then when the voltage pressure is taken off the capacitor by reversal of the ac cycle, the capacitor gives the energy back to the circuit as a spring releases energy when the compressing force is removed. From this it can be seen that a capacitor will allow current flow in an ac circuit by "taking in" energy and "returning" it during successive half cycles. But because a capacitor contains an insulating material between its plates, it will not allow direct current to flow.

The property of a capacitor to allow

PRINCIPLE OF OPERATION

If battery is connected to a capacitor, current will flow to one plate and away from the other, leaving one more positive. Current then stops; current will not flow through dielectric.

Dielectric   Plates

Battery

If battery is removed, plates remain in this condition; one is more positive than the other. Capacitor is said to be charged.

If leads are connected together, current will flow from one plate to the other until both are again neutral and current stops. Capacitor is said to be discharged.

If battery is reconnected in opposite direction, current will again flow momentarily and the other plate will receive the positive charge.

Removal of voltage by connecting leads again causes momentary current flow, discharging capacitor.

Thus, current flows only when voltage is changed. In an alternating current circuit, voltage is constantly changing; therefore current flows continuously.

INDUSTRIAL CAPACITORS FOR POWER-FACTOR CORRECTION

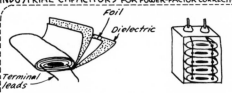

Foil

Dielectric

Terminal leads

Capacitors are connected in parallel to form a cell, stacked in a can, and covered with an insulating compound.

Cells are assembled in housing to form larger unit of desired capacity.

Units are mounted in racks for connection to power system.

SMALLER CAPACITORS

CERAMIC:

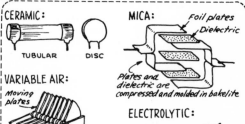

TUBULAR   DISC

MICA: Foil plates
                 Dielectric

VARIABLE AIR:

Moving plates

Fixed plates

Plates and dielectric are compressed and molded in bakelite

ELECTROLYTIC:

COLOR CODE for mica capacitors identifies capacitor's rating in micromicrofarads:

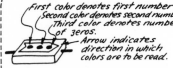

First color denotes first number
Second color denotes second number
Third color denotes number of zeros.
Arrow indicates direction in which colors are to be read.

Example: RED-BLACK-RED indicates 2-0-00, or 2000 micromicrofarads, or .002 microfarads.

CODE	
Black	0
Brown	1
Red	2
Orange	3
Yellow	4
Green	5
Blue	6
Violet	7
Gray	8
White	9

CAPACITOR SYMBOLS

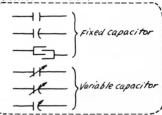

Fixed capacitor

Variable capacitor

current to flow when the voltage across it changes is called "capacitance." This property is measured in units called "farads." Capacitors used in electronic circuits and motors are generally measured in microfarads (millionths of a farad) and micromicrofarads. Power capacitors used for power factor correction are rated in kilovars (*kilovolt-amperes reactive*), a rating which more directly indicates the ability of a particular unit to improve power factor.

Power Capacitors

The capacitors used to improve power factor in industrial plants and in other power systems are commonly called "power capacitors." In such units, two materials are used to provide the dielectric: thin kraft paper and an impregnating dielectric liquid. The paper is impregnated with this liquid which is a nonflammable compound known as askarel and is placed between the electrodes. The combination provides safe application at the higher voltages, increased stability for long life and greater capacity for power factor correction.

The plates in power capacitors are very thin sheets of aluminum foil. Sheets of foil between sheets of impregnated kraft paper are wound into rolls which are provided with terminal strips and then bound into bundles and connected in parallel within the housing. Number and size of rolled packs depends upon the voltage and capacity rating of the unit.

Power capacitors are equipped with internal resistors connected across the terminals of the rolled electrode packs. Such a resistor provides for draining off the voltage charge of a capacitor when it is disconnected from the line, eliminating shock hazard to personnel.

Both single- and three-phase capacitors are available for power factor correction service. Voltage ratings range from 230 volts to primary voltages, in typical ratings of 0.5 kilovar, 5 kilovar and 25 kilovar.

Application of power capacitors to improve power factor of a feeder, sub-feeder or circuit is based on the characteristics of alternating-current and the effects of load. Inductive loads on ac circuits—motors, fluorescent lighting, electromagnetic loads and any device using magnetic coils—draw more current from the line than they actually convert to power. The greater the ratio of extra current to total current, the lower the power factor. The extra current is called "magnetizing current" and is necessary to set up required magnetic fields. By using a capacitor close to the load and connected in the circuit, the capacitor will provide the magnetizing current, relieving the circuit of this current. The results of this action include: improved voltage stability, released system capacity and lower power bills.

Smaller Capacitors

- **Variable capacitors** are used where the value of capacity in the circuit must be continuously variable, as in radio equipment. Usually, these consist of two sets of rotating metal plates which mesh with each other with air serving as the dielectric.
- **Paper capacitors** use tinfoil sheets, insulated with waxed paper as the dielectric and rolled together into a small cylinder, with a terminal lead coming out of each end. These are used in electronic circuits.
- **Electrolytic capacitors** derive their action from a chemically formed dielectric film within the unit. These capacitors are cylindrical in shape and are available with several types of terminal arrangements. Electrolytic capacitors are extensively used in all types of electronic power supplies and circuits and are the kind of capacitors used in the many types of capacitor-start and capacitor-run motors. Both polarized and non-polarized electrolytic capacitors are made.
- **Mica and ceramic capacitors** are other types widely used in electronic circuits. They use mica and ceramic, respectively, as the dielectric.

Batteries

A N ELECTRIC battery is a device used to produce electrical energy from chemical energy.

Essentially, all batteries consist of two dissimilar electrodes immersed in a conducting solution called an electrolyte. Chemical action between the electrodes and the electrolyte causes current to flow from one electrode to the other through the electrolyte when the electrodes are connected to an external load such as a lamp bulb. This basic arrangement is called a cell; a battery may consist of one or more cells.

• **Rating**—Batteries are rated in ampere-hours. A battery rated at 120 ampere-hours will furnish roughly 10 amperes for 12 hours, 5 amperes for 24 hours, 20 amperes for 6 hours, etc., before it is exhausted. (Actually these combinations are not strictly attained in practice; they vary according to the amount of current drawn. Constant use at heavy currents results in shorter-than-rated life, while relatively small current drain lengthens the rated life.)

• **Classification**—Batteries are classified as either primary or secondary. The primary battery contains an electrode which is consumed as the battery is discharged; thus this type of battery is either discarded eventually or the electrode and the electrolyte are re-placed to restore the battery to operating condition. The electrodes and electrolyte of a secondary battery also change chemically as the battery is discharged; however the battery may be restored to its original condition without replacing the elements by sending dc current through it in the opposite direction. The process restores the potential and is known as "charging" the battery.

Primary Batteries

Primary batteries in use today are either of the "wet" or "dry" variety, these terms referring to the relative condition of the electrolyte. Primary batteries are used as voltage standards for laboratory use, railroad signal circuits, fire and burglar alarms, telephone and telegraph systems, and intercom and annunciator systems. The familiar dry cell is a form of primary battery which is used extensively for signal and communications systems.

The principal types of wet primary batteries in use today, together with their electrodes and electrolytes, are as follows: Daniell's cell (copper, zinc, copper sulphate); gravity cell (copper, zinc, copper sulphate); Fuller cell (carbon, zinc, potassium bichromate plus sulphuric acid); Lalande cell (copper, zinc, caustic soda); Laclanche cell (carbon, zinc, sal ammo-

No. 6
Cell

Flashlight
(Type D)
cell

CYLINDRICAL DRY CELLS

Commonly used in flashlights and other instruments. One connection is made at cell top, the other at the base — except for No. 6 cell, which has two top terminals.

COMMON CYLINDRICAL-TYPE DRY CELLS with nominal dimensions (inches):

ASA Designation	Diameter	Height of Can
N	27/64	1 1/64
N	7/16	1 3/32
R	17/32	1 3/8
AAA	25/64	1 11/16
AA	17/32	1 7/8
A	5/8	1 7/8
B	3/4	2 1/8
C	15/16	1 13/16
CD	1	3 3/16
D	1 1/4	2 1/4
E	1 1/4	2 7/8
F	1 1/4	3 7/8
G	1 1/4	3 13/16
6	2 1/2	6

WAFER-TYPE DRY CELLS

New type cell uses flat carbon and zinc electrodes with electrolyte of artificial manganese dioxide. Spot of silver wax provides improved contact between cells when stacked. Entire stack is enclosed in thin plastic film and encased to form larger battery.

Wafer-type
Cell

FLAT-TYPE DRY CELLS

Carbon and zinc plates are separated by sheet of paper dipped in electrolyte. Cells are stacked and enclosed to form larger battery. Flat cells make more efficient use of space than round type.

Flat-type
cell

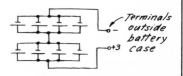

Terminals outside battery case

DRY BATTERY may use eight 1½-volt cells arranged in series-parallel. Four cells in parallel give 1½ volts at four times the capacity of a single cell; the two groups in series make 3 volts available at terminals.

Cells interconnected with straps

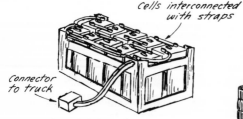

Connector to truck

STORAGE BATTERIES used for industrial motive power are generally lead-acid or nickel-iron type. Batteries are permanently mounted in steel trays to fit lift trucks, hand trucks, tractors, mine locomotives, etc. Entire tray is removed from vehicle for recharging.

Mercuric-oxide cells

MERCURY-TYPE BATTERY is recent development. Used as efficient source of dc power for alarm systems, hearing aids and instruments, it has a zinc anode, mercuric oxide electrode, and potassium hydroxide electrolyte. Typical open-circuit voltage is 1.34 volts, kept constant by trickle-charging.

niac). A glass jar is characteristic of the wet primary battery.

The dry cell employs a carbon rod as a positive electrode, a cylindrical zinc container as the negative electrode, and a solution of sal ammoniac and zinc chloride as the electrolyte within the container.

The cell is described as "dry" because the space between the electrodes is not filled with a liquid but contains an absorbent lining which is saturated with the electrolyte. The top of the cell is sealed; thus it is thrown away when it becomes exhausted.

Secondary Batteries

Secondary batteries are commonly known as storage batteries. They are similar in operation to the primary batteries discussed; however the materials used are such that they may be restored through charging. Types of storage batteries in common use are the lead-acid, nickel-iron-alkaline, nickel-cadmium-alkaline, silver-zinc-alkaline, and silver-cadmium-alkaline.

The lead-acid battery consists of one or more positive electrodes (anodes) of lead peroxide, an electrolyte solution of sulphuric acid and water, and one or more negative electrodes (cathodes) of sponge lead. All anodes are connected together inside the battery, as all cathodes. Each electrode is in the form of a grid with a paste in the active ingredient of the electrode.

These batteries contain a "sediment space" at the bottom to take care of a gradual accumulation of sediment from the active ingredients of the electrodes which forms on the bottom during normal operation. Water is gradually lost during charging and by normal evaporation, hence additional water must be added from time to time to maintain the required amount of electrolytic solution.

A typical alkaline electrolyte is a solution of caustic potash and lithium hydrate. It does not enter into chemical action with the electrode material as does the acid of the lead-acid battery, but merely acts as a medium for the conduction of current between the two electrodes.

The choice of an acid or alkaline type depends upon the use to which the battery is to be put as well as economic considerations. The alkaline type has a higher first cost, but it is rugged, light in weight, has a long life, and maintenance costs are low.

Its internal resistance, however, is higher than that of the lead-acid type. The expected full-charge voltage of a single lead-acid cell is usually taken as 2 volts, that of the alkaline type 1.2 volts. Voltages desired are attained by connecting single cells in series.

Applications of storage batteries include (in addition to certain of the single systems mentioned under primary batteries above) standby power and control, stationary engine cranking, switchgear closing and tripping, motive power, vehicular starting, and emergency lighting.

Battery-operated power plants have long been used on farms for regular or standby use. Large buildings having diesel emergency power units frequently use cranking batteries to turn over the engine when the normal building power supply fails. Others having no engine-generator use banks of batteries to furnish electrical energy for necessary lighting and power in emergencies. Power stations and substations frequently depend upon storage batteries to provide power for the operation of switchgear and other control functions when all other sources have failed. Most manufacturing plants and warehouses maintain electric materials-handling trucks powered by storage batteries.

The materials used in constructing storage batteries are governed by the intended use. Batteries used for ignition work or lighting systems usually employ a hard rubber container; heavier-duty applications such as industrial lift-truck powering require a more durable metal-clad case. Stationary installations such as emergency standby systems may use glass jars.

Battery Chargers

A BATTERY CHARGER is an electrically powered unit designed to re-establish the chemical characteristics which give a charge to a secondary battery.

When a battery discharges, the chemical reaction produces an electric current flow through the circuit of the load connected to the battery, in response to the potential charge of the battery.

When the chemical reaction has been exhausted (established equilibrium in the electrolytic process), a potential charge no longer exists and current no longer flows. But in secondary or storage batteries, an application of a dc potential to the battery terminals with polarity of connections to produce current flow opposite to that when the battery acts as a source of power, will reverse the chemical reactions, reversing the condition of the battery from "discharged" to "charged." Battery chargers provide the dc potential source and necessary controls for accomplishing this action.

Arrangements for battery recharging vary with the application of the batteries. For instance, storage batteries used as motive power for industrial materials-handling trucks are commonly recharged in the plant's battery shop where a permanent installation of dc charging circuits makes available a substantial capacity for recharging a large number of batteries at one time. Such layouts generally consist of a conversion assembly for changing alternating current power from the plant electrical system into direct current power for charging, along with extensive supply conductors for connecting to the battery loads and associated controls.

Batteries used for emergency lighting in buildings or for powering electrically-operated switchgear are also provided with a permanent charging arrangement at the location of the battery racks. However, in such a case, the charger is generally a small assembly which operates only to keep the batteries charged up to their normal potential. Relays and other controls provide for taking the battery power for its intended purposes.

Still another type of charging arrangement is that provided by portable unit chargers. Such units can be moved to the battery location.

This type of charger is usually used for charging only one battery at a time. A familiar example of the portable type of charger is the charger used in automobile service stations.

A storage battery converts chemical energy to electrical energy when connected to an external load. The chemical energy is restored by charging; electrical energy supplied by the charger is converted back to chemical energy. This transfer of energy is shown by the following diagrams, assuming a battery with a sponge lead negative electrode, a lead peroxide positive electrode, and sulphuric acid electrolyte.

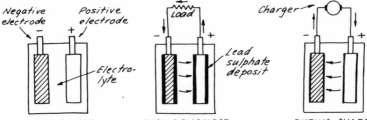

FULLY CHARGED	DURING DISCHARGE	DURING CHARGE
The positive and negative electrodes are totally different in chemical composition, causing a potential difference to exist between them.	Deposit of lead sulphate is created on electrodes. Eventually both are virtually alike, with almost no voltage difference between them.	Charger feeds current back into positive electrode, breaking up lead sulphate deposits. Electrodes finally return to their original chemical composition.

In all cases, however, it is the function of the battery charger to bring the batteries up to their rated charge—without overcharging, overheating or excessive gassing. To do this, the charger must have a dc voltage output equal to the battery voltage where batteries are charged in parallel or to the sum of the battery voltages where the battery supply consists of a series circuit of a number of storage batteries, such as a 120-volt battery supply for emergency lighting.

Basically, there are two types of battery chargers: the motor-generator set type and the rectifier type.

M-G Chargers

Battery charging using m-g sets is common in industrial applications where a large number of batteries are used—either for recharging motive power truck batteries or maintaining a charge on a series connection of batteries used for emergency lighting or for electrically operated switches or circuit breakers. In such applications, an ac circuit is connected to the drive motor which is mechanically coupled to the shaft of a direct current generator. The motor and generator are mounted on a common base.

Associated controls installed with m-g chargers provide regulation of the "charging rate" of current flow into the batteries. In most cases, this control is automatic and adjustable according to the type of battery being recharged and the condition of discharge.

Typical m-g set chargers are made for lead-acid and nickel-alkaline batteries. They are rated according to the ac supply voltage (220, 440, 550 3-phase, etc.) hp of the driving motor and kw of the generator. Manufacturers commonly relate the charging capability of various models to the number of cells and the ampere-hour capacities of batteries, based on the maximum number of hours required to charge batteries. Assemblies usually include motor starter and protection for the induction drive motor. Both single-circuit and multi-circuit

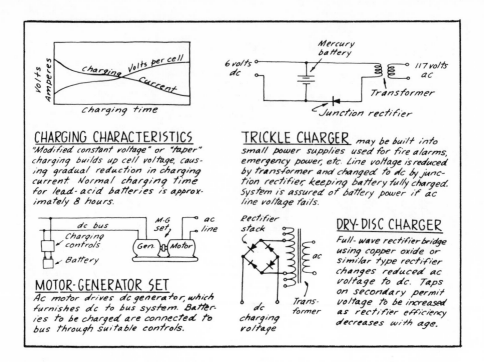

CHARGING CHARACTERISTICS

"Modified constant voltage" or "taper" charging builds up cell voltage, causing gradual reduction in charging current. Normal charging time for lead-acid batteries is approximately 8 hours.

TRICKLE CHARGER *may be built into*

small power supplies used for fire alarms, emergency power, etc. Line voltage is reduced by transformer and changed to dc by junction rectifier, keeping battery fully charged. System is assured of battery power if ac line voltage fails.

MOTOR-GENERATOR SET

Ac motor drives dc generator, which furnishes dc to bus system. Batteries to be charged are connected to bus through suitable controls.

DRY-DISC CHARGER

Full-wave rectifier bridge using copper oxide or similar type rectifier changes reduced ac voltage to dc. Taps on secondary permit voltage to be increased as rectifier efficiency decreases with age.

outputs are available, for charging any number of batteries.

Static Converters

Rectifier type battery chargers use a completely static—no moving parts—method for converting from ac to dc. In such units, the ac input is connected to a configuration of rectifying elements which pass current in one direction but not in the other. This dc output is then connected to the battery terminals.

The most common type of rectifier charger is the dry disc type which uses a semi-conductor material for rectification. The most common disc is that made of selenium, although germanium and silicon have lately found application in such rectifier circuits. Disc-type rectifiers offer very efficient ac to dc conversion (very little energy is lost in the conversion process itself), and easy maintenance.

Typical selenium chargers are made in a variety of sizes and capacities,

suited to all types of battery charging applications, including: industrial truck batteries, batteries for switchgear operation, batteries for mine locomotives, batteries for emergency power use, vehicular starting batteries. These units are rated according to ac supply voltage, input power requirements, output voltage and current, and characteristics suited to particular applications. They are also rated in amp-hour capacities for specific numbers of cells, for given time of charging.

Depending upon the output voltage required, selenium chargers are also equipped with necessary transformers. Some units are convection cooled; others, fan cooled or oil cooled. The complete assembly is contained within a steel enclosure, with external controls and meters.

Another type of rectifier charger uses electronic tubes for the rectifying element. These units are made in a number of sizes for a range of charging applications.

Grounding
Accessories

GROUNDING in electrical systems refers to the intentional means by which the circuits and/or the enclosures of electrical equipment (frames, housings, cabinets, conduits and other raceways) are maintained at ground potential.

Concept of Grounding

Skillful application of grounding accessories can be assured by a thorough understanding of grounding itself. The more important facts about grounding are as follows:

1. Grounding is one form of protection against the potential hazards of electrical application. Other forms of protection which are suitable alternatives to grounding under certain conditions are insulation and guarding.

2. In electrical circuits, grounding of one of the conductors limits the maximum potential to ground from the other conductors to ground under normal conditions. It also limits voltage which might otherwise occur through exposure to lightning or other voltages higher than that for which the circuit is designed.

3. Exposed conductive materials enclosing wires and cables and electric equipment and devices are grounded for the purpose of preventing a potential above ground on such enclosures. In the case of a grounded electrical system (where a circuit conductor is grounded in addition to the grounding of the conduits, boxes, frames, etc.), if one of the ungrounded circuit conductors were accidentally to come in contact with the conduit or a motor frame, for instance, the circuit fuse or circuit breaker would open the circuit. In the case of an ungrounded system (where no circuit conductor is grounded), if one of the ungrounded conductors were accidentally to come in contact with a metallic enclosure, there would be no hazardous potential on the enclosure because it is already at ground potential.

In either type of system, if the metallic enclosures were not grounded and one of the ungrounded system conductors came in accidental contact with the enclosure, there would be a dangerous potential to ground from the enclosure. Anyone touching the enclosure and simultaneously making contact with a grounded surface or object (the earth itself, a basement floor, a water pipe, a radiator, building steel, etc.) would get a shock —ranging from slight to fatal depending upon the voltage and the resistance of the path from the hot conductor to ground.

For the above reasons, most electric circuits and enclosures are grounded. Circuits are grounded at services and at other sources such as generators or transformer secondaries. When an interior wiring system is grounded at the service entrance, it is not grounded elsewhere. The interconnected system of conduits and enclosures is also grounded at the service entrance and must be made com-

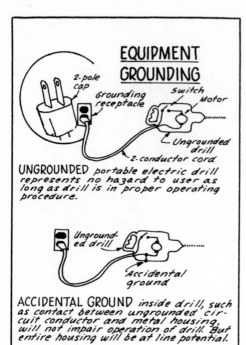

EQUIPMENT GROUNDING

2-pole cap

Grounding receptacle

Switch

Motor

Ungrounded drill

2-conductor cord

UNGROUNDED *portable electric drill represents no hazard to user as long as drill is in proper operating procedure.*

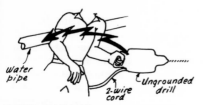

Unground-ed drill

Accidental ground

ACCIDENTAL GROUND *inside drill, such as contact between ungrounded circuit conductor and metal housing, will not impair operation of drill. But entire housing will be at line potential.*

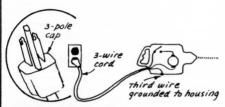

Water pipe

2-wire cord

Ungrounded drill

SHOCK HAZARD *to user exists with this condition, since contact with a grounded object such as a water pipe will cause current to flow from drill housing through user to pipe.*

3-pole cap

3-wire cord

third wire grounded to housing

GROUNDING *drill housing to circuit ground by means of a third grounding conductor and 3-pole cap provides ground potential on housing. Now any accidental contact between housing and ungrounded circuit conductor will cause fuse to blow and render drill inoperable, eliminating shock hazard.*

mon (bonded) with the grounded circuit conductor.

The use of grounding accessories to accomplish the above objectives is closely tied to grounding requirements of the National Electrical Code. Article 250 of the code details the minimum requirements for safety.

Typical Devices

The following are the common types of accessories used for grounding:

Grounding bushing—This is a special form of conduit bushing which is equipped with a side-mounted mechanical clamp for connecting a bonding jumper. Such bushings are used where service conduit connects to meter boxes or to service equipment. The bonding jumper assures electrical continuity of the grounding circuit for the service layout. Such jumpers must be used around concentric knockouts in the service panel or meter box. But where concentric knockouts do not exist, the conduit can be bonded to the box or panel enclosure by means of **bonding bushings,** bonding locknuts or bonding wedges.

Bonding bushing—This is somewhat similar to the grounding bushing except that a screw is arranged to be driven through the bushing body to "bite" into the wall of the enclosure to which it is desired to "bond" the conduit. The screw may pass through the bushing perpendicular to the box wall or at an angle to wedge between the conduit and the box wall. Such a bushing may also provide for connecting a jumper wire to ground.

Bonding locknut—This is similar to a standard locknut, with an angle screw through the body to provide a bonding "bite" into the box wall, as described above. This locknut is used on the inside of the service equipment, with a standard locknut outside and a regular bushing (fully insulating type, for instance) over the bonding locknut Such locknuts are used in place of bonding bushings.

Bonding wedge—This is a flat C-shaped device designed to be used between the wall of the box and a regular bushing. It is equipped with "biting" screws on its side to provide a bond between box and conduit.

Grounding clamp—This is a device which provides for connection of a grounding conductor to the water pipe or ground rod. Made in a range of sizes, typical clamps provide for connection of bare or insulated grounding conductors, armored ground wires and grounding conductors in conduit. Clamps are made to fit various sizes of water pipes, and special units are made for connecting to ground rods.

Ground rods—These are made electrodes for use where a satisfactory water piping ground is not available. Typical rods are made in lengths up to 40 feet, and in diameters varying from ⅜-in. to 1 in. They are made in single lengths or in sections. Some rods have molten-welded copper exteriors over steel cores; others are hot-dipped galvanized. Rods may be hand driven or power driven.

Ground conductors — Common grounding wires include: bare armored ground wire, with either a solid or stranded conductor, in sizes from No. 8 to No. 4; bare copper or aluminum conductors, either solid or stranded; and insulated conductors. According to the code, system or common grounding conductors may be of copper or other corrosion-resistant material—solid or stranded, bare or insulated. In addition to the foregoing type of conductor, a bus-bar, a rigid conduit, a steel pipe, EMT or armor on cable may be used for grounding equipment and conductor enclosures. Galvanized steel ground wire is used for pole-line grounding.

Ground-wire moulding—This is wood molding made with a C-shaped cross section for covering and mechanically protecting bare ground wires run along walls or down poles.

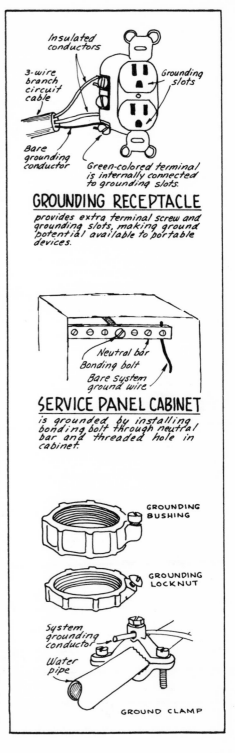

GROUNDING RECEPTACLE
provides extra terminal screw and grounding slots, making ground potential available to portable devices.

SERVICE PANEL CABINET
is grounded by installing bonding bolt through neutral bar and threaded hole in cabinet.

GROUNDING BUSHING

GROUNDING LOCKNUT

GROUND CLAMP

Switches

A S USED in the electrical industry, the term "switch" refers to a device for making, breaking or changing the connections in an electrical circuit, under a certain rated load but not under short - circuited conditions. Because switches are so widely and generally used, there are many types and sizes of switches suited to particular conditions of application. The following covers construction, operation and application of the more common switches in use today.

Classification

Construction and operating characteristics provide the major basis for classifying switches. These characteristics are broken down as follows:

• **Number of Poles**—In any switch, there are one or more terminals to which incoming wires are connected and the same number of terminals to which the outgoing wires are connected. These terminals are called "poles." The number of terminals for incoming wires (or the number of terminals for outgoing wires)

determines how a switch will be described according to number of poles. A single-pole switch, for instance, will have one terminal for the wire coming in and one terminal for the wire going out. When the switch is closed, it forms part of the current path through the wire. When the switch opens, it breaks the current path through the wire. A two- or double-pole switch provides the same action for two wires passing through the switch. A three- or triple-pole switch can close or open the current path through three wires; a four-pole switch, through four wires; and a five-pole switch, through five wires.

In many cases, especially where the wiring system has a grounded neutral conductor, the neutral connection must not be broken as the wires pass through the switch. For these cases, there are two ways to carry the neutral through the switch. First, a switch with a number of poles equal to one less than the number of wires can be used; the neutral wire can just be carried directly through the box and the other wires connected to the switch poles. The second way is to use a "solid-neutral" switch. This is a switch with a number of poles equal to the number of wires carried through the switch, but with a solid conductor strap between one of the incoming poles and the corresponding outgoing pole. The neutral wire is connected to these poles. Opening the switch opens the other wires but does not open the neutral.

● **Number of Throws**—A switch may have one, two or more closed positions. If a switch can be closed by moving the contact to only one position, the switch can be closed in either of two positions, it is called "double-throw." There are also multiple-throw switches, such as rotary or dial switches, which can be closed in several positions.

● **Contact**—There are several types of contact elements in switches. One is the knife blade used in many switches. In such switches, flat knife-like blades are connected to the outgoing poles and are pivoted to allow insertion and removal of the bladed form contact pressure jaws which are connected to the poles on the incoming side of the switch. The knife-switch principle is used in both open and enclosed units.

Another contact arrangement is the "butt-contact" type in which movable contacts which are connected to the outgoing poles or load side of the switch are moved into a high-pressure butt with stationary contacts on the line side of the switch.

In certain low-capacity and control circuits, a small switch is often used which has mercury for the contact element. Such a switch is generally a small glass tube with a stationary contact at each end and an amount of mercury in the tube. Mercury switches may be operated by a handle or automatically by a movable device like a thermostat. In the "off" position, the mercury is at one end of the tube touching only one contact. When the switch is moved to the "on" position, by tilting the tube, the mercury spreads the length of the tube and provides an electrically con-

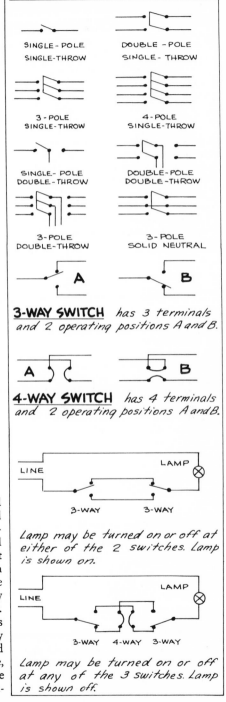

COMMON SWITCH SYMBOLS

SINGLE-POLE SINGLE-THROW

DOUBLE-POLE SINGLE-THROW

3-POLE SINGLE-THROW

4-POLE SINGLE-THROW

SINGLE-POLE DOUBLE-THROW

DOUBLE-POLE DOUBLE-THROW

3-POLE DOUBLE-THROW

3-POLE SOLID NEUTRAL

3-WAY SWITCH *has 3 terminals and 2 operating positions A and B.*

4-WAY SWITCH *has 4 terminals and 2 operating positions A and B.*

LINE   LAMP
3-WAY   3-WAY

Lamp may be turned on or off at either of the 2 switches. Lamp is shown on.

LINE   LAMP
3-WAY   4-WAY   3-WAY

Lamp may be turned on or off at any of the 3 switches. Lamp is shown off.

MAGNETIC SWITCH

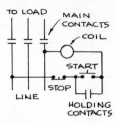

Pressing start button ener-gizes coil, closing main con-tacts and holding contacts. Holding contacts keep coil energized after button is released. Pressing stop button de-energizes coil, opening all contacts and cutting off power from load.

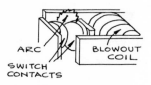

Magnetic blowout coil is oft-en provided as part of magnetic switch to pre-vent burning of contacts by arc when contacts open. Magnetic field of coil draws arc upward and away from contacts.

SAFETY SWITCH

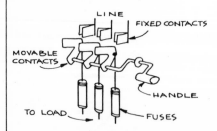

Safety switch consists of oper-ating handle, movable and fixed contacts and fuses, all except handle enclosed in cabinet.

ductive path between the contacts in the tube.

- **Number of Breaks**—When opening a wire, a switch may break contact at one or two points. If a switch breaks contact between an incoming pole and the corresponding outgoing pole at only one point, the switch is called a "single-break" type. If it breaks contact at two points, it is called a "double-break" type. A double break of contact better suppresses the arc which is developed.

- **Type of Contact Break** — Most switches provide for breaking a circuit in the air around the switch contacts. These are called "air-break" switches. Another widely used type of switch is the "oil-switch" which has its contact parts immersed in oil, the breaking of the contact taking place in the oil.

- **Method of Operation** — Switches may be operated (opened and closed) by several methods: manually; electromagnetically; by motor drive; or automatically by some external device. Manual switches are by far the most common type of switch. They generally operate by some form of lever action, although there are many manual switches of the rotating or drum type.

Magnetic switches are widely used in motor control work and where it is necessary to operate the switch contacts from a remote pushbutton station.

There are many types of manual switches used for such power applications as: motor disconnect switches, service entrance switches, safety switches, and feeder disconnect switches. Other types of manual switches are those used for branch circuit and low capacity switching. These include; pushbutton wall switches, toggle or tumbler switches and rotary snap action switches.

A dial switch has a contact mounted on a rotating arm which can bring the arm contact against any one of several fixed contacts arranged in circular form. This type of switch is used for control and instrument applications and for tap-changing in transformers. A drum switch is also a rotating type multi-contact switch used for motor starting and control applications.

SWITCHES are generally mounted in enclosures to protect an operator's hand from contact with current-carrying parts and to protect the switch mechanism from damage. In some cases, when only authorized personnel have access to switches, they may be mounted without an enclosure. This is frequently the case with switchboards and panelboards, although even in this application, the exposure of live parts of the switch should be avoided. A switchboard on which current-carrying parts of switches are exposed is commonly called a "live-front" switchboard. A "dead-front" switchboard has no live parts exposed to possible accidental contact by personnel. Of course, when a switch is enclosed, it should be operable with the enclosure closed; i.e., it should have an external handle or button.

Enclosure Types

Because switches are basic electrical devices and their application runs the gamut of electrical systems, they are available with a wide range of enclosure types, each suited to a particular set of job conditions. As set forth by the National Electrical Manufacturers Association, standard types of switch enclosures are as follows:

Type I, General Purpose—for indoor applications where atmospheric conditions are normal.

Type II, Driptight—has shields to protect against entry of moisture into the enclosure when used in areas where condensation is severe.

Type III, Weather Resistant—heavy duty enclosure which provides protection of the switch against rain and snow in outdoor applications.

Type IV, Watertight—constructed to keep water out of the enclosure when the unit is subjected to a stream

TUMBLER SWITCH

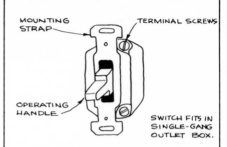

MOUNTING STRAP

TERMINAL SCREWS

OPERATING HANDLE

SWITCH FITS IN SINGLE-GANG OUTLET BOX.

The flush-type tumbler switch, used extensively in interior wiring for control of lights and small appliances, is also the design often used for three- and four-way switches and mercury switches.

MERCURY SWITCH

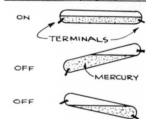

ON

TERMINALS

OFF

MERCURY

OFF

Mercury switch consists of tube containing mercury, two contacts, and mechanism for tilting tube.

KNIFE SWITCH

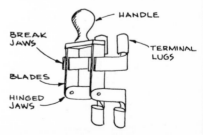

HANDLE

BREAK JAWS

TERMINAL LUGS

BLADES

HINGED JAWS

Movable copper blades are forced between spring-type forked jaws to close circuit.

of water, as it might be in dock-side and other wet area applications.

Type V, Dust-tight—a gasketed or otherwise protected enclosure to keep dust out, as in grain elevators, cement mills, etc.

Type VI, Submersible—constructed to allow operation of the switch with the enclosure under water, as in manholes, mines, etc.

Type VII, Hazardous Location—an enclosure for an air-break switch, meeting requirements of the National Electrical Code for Class I, Group D hazardous locations.

Type VIII, Hazardous Location—an enclosure for an oil-immersed switch, meeting requirements of the National Electrical Code for Class I, Group D locations.

Type IX, Hazardous Location—an enclosure meeting requirements of the National Electrical Code for Class II hazardous locations.

Type X, Bureau of Mines—especially constructed for use in coal mines.

Types of Switches

• **General-purpose switches** are used for general distribution and circuit applications. The enclosed type of general purpose switch is called a "safety switch." They are rated according to the severity of operating conditions under which they may be used. A Type A switch is designed for heaviest duty operation, allowing high frequency of operation and heavy current-carrying capacity. Type C general-purpose switches are not constructed as heavily as Type A switches and are limited to applications where load conditions and overloads are not severe and where the switch is not operated frequently. Type D switches are for light loads and where operation is very infrequent, such as service entrance applications.

• **Safety switches** are made in a variety of sizes and models, with the handle on the side or front of the enclosure. Safety switches are made with and without clips for including fuses in the enclosure.

• **Disconnecting switches** are actually isolating switches used to isolate a circuit from its source of power. They can be opened only after the circuit has been broken by some other switching device; they cannot interrupt the circuit current. Such switches are generally used to isolate motors for repair or inspection, and they can often be locked in the open position. There are both indoor and outdoor, open or enclosed disconnect switches.

• **Motor circuit switches** are rated in horsepower to match the rating of the motor circuit on which they are used. They must be able to interrupt the maximum overload current of a motor for which they are rated. These switches are simply general-purpose switches with horsepower rating for use on motor circuits.

• **Motor-starting switches** are used to close the circuit to motors which can be connected directly across the line. Although regular motor circuit switches can often be used for such duty, better practice dictates the use of switches designed for motor starting. Motor starting switches may or may not incorporate a protective device to open the circuit to the motor upon overload.

• **Field switches** take their name from the circuits in which they are used—field circuits of generators and synchronous motors.

• **A service entrance switch** is simply a fused safety switch used in the service entrance conductors at the point at which the conductors come onto the premises. There are many types of service switches, including: fused safety switch, fused main switch and fused range switch with branch circuit fuses in one enclosure, called "service centers"; and types with fused safety switch and plug fuse receptacles in

DISCONNECTING SWITCH, single pole, single throw.

Type often used in substations to isolate high-voltage conductors from substation equipment. Switches are opened from below, phase by phase, by use of hookstick held by operator.

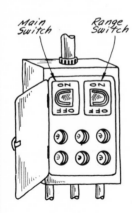

To Utility

Hookstick

To Transformer

FIELD DISCHARGE SWITCH has extra contact connected to a resistance. When switch is opened, one blade makes with extra contact as it leaves main contact, connecting resistance across generator field. This prevents high voltages from being induced in field when circuit is suddenly opened. (Dotted lines indicate connections to switch contacts.)

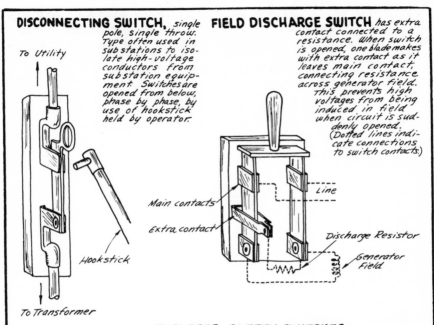

Main contacts

Extra contact

Line

Discharge Resistor

Generator Field

ENCLOSED SAFETY SWITCHES may contain such features as arc-quenching, interlocked cover, and spring-operated quick-make-and-break action. They are used for such applications as motor loads, lighting and appliance loads, and service entrances.

SERVICE SWITCH, fuse puller type, may contain also a range switch plus plug fuse receptacles for branch circuits.

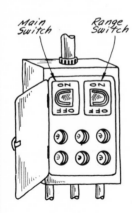

Main Switch

Range Switch

Main switch and range switch are placed in OFF position by pulling fuse unit out of enclosure by means of handle, and reinserting upside down.

WEATHER - RESISTANT
Side-Operated

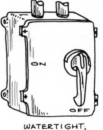

WATERTIGHT,
DUST-TIGHT
Front-Operated

Enclosed switches usually include provision for locking handle in OFF position. Where more than one electrician is working on the same circuit, each can put on his own lock. The switch cannot be closed until all locks have been removed.

GENERAL PURPOSE
Front-Operated

175

one enclosure. Service entrance switches are made for both indoor and outdoor applications.

The wide range of small switches used for lighting fixtures, oil burners, fans and similar utilization devices on branch circuits are called "wiring switches." These are generally snap-action toggle switches operated by a lever or rotary type handle. Mercury switches, although operated by a lever handle, are not snap-action switches. Wiring switches are compact units which mount readily in a switch or outlet box. Among the commonly used wiring switches are combination units which consist of several switches, receptacles and/or pilot lights.

Relay Switching

IN MODERN electrical systems for all types of buildings, there is a growing trend toward the convenience, flexibility and economy of various types of remote control switching of lighting and appliance loads.

One type of remote control switching makes use of small relays which have contacts to open and close the circuit to the lighting load and have a small electromagnet operating at some voltage under 50 volts to control the opening and closing of the contacts. This type of control is commonly called "remote control" switching, "low voltage" switching, or "relay" switching.

Application

The use of low voltage relay switching offers many advantages for use in residences, commercial and institutional buildings and in industrial plants.

It can be used for controlling one or any number of outlets on a circuit or for controlling the entire load connected to the circuit. It offers greatly reduced wiring requirements for control of lighting circuits compared to direct manual switching at the voltage of the circuit.

If one or more lighting fixtures is to be switched at line voltage by a typical wall switch, the wiring to the switch and the switch itself must be so constructed and installed as to meet the requirements of regular circuit wiring.

If the circuit voltage is 277-v instead of 120-v—as is being found in more and more modern installations—the wiring to the switch must not only be standard wire-and-conduit or approved cable construction, but the

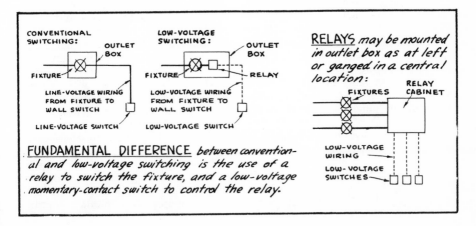

CONVENTIONAL SWITCHING:
OUTLET BOX
FIXTURE
LINE-VOLTAGE WIRING FROM FIXTURE TO WALL SWITCH
LINE-VOLTAGE SWITCH

LOW-VOLTAGE SWITCHING:
OUTLET BOX
FIXTURE
RELAY
LOW-VOLTAGE WIRING FROM FIXTURE TO WALL SWITCH
LOW-VOLTAGE SWITCH

RELAYS may be mounted in outlet box as at left or ganged in a central location:
FIXTURES
RELAY CABINET
LOW-VOLTAGE WIRING
LOW-VOLTAGE SWITCHES

FUNDAMENTAL DIFFERENCE between conventional and low-voltage switching is the use of a relay to switch the fixture, and a low-voltage momentary-contact switch to control the relay.

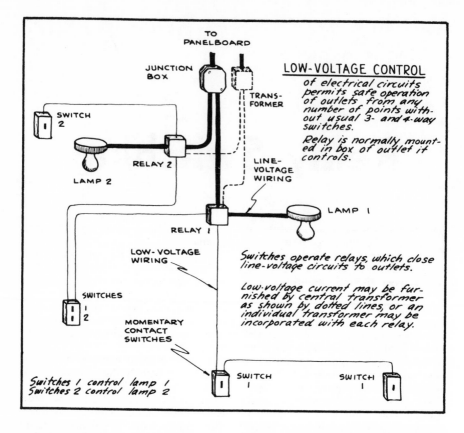

TO PANELBOARD

JUNCTION BOX

SWITCH 2

RELAY 2

LAMP 2

TRANS-FORMER

LINE-VOLTAGE WIRING

LOW-VOLTAGE CONTROL

of electrical circuits permits safe operation of outlets from any number of points without usual 3- and 4-way switches.

Relay is normally mounted in box of outlet it controls.

LAMP 1

RELAY 1

LOW-VOLTAGE WIRING

Switches operate relays, which close line-voltage circuits to outlets.

Low-voltage current may be furnished by central transformer as shown by dotted lines, or an individual transformer may be incorporated with each relay.

SWITCHES 1 2

MOMENTARY CONTACT SWITCHES

Switches 1 control lamp 1
Switches 2 control lamp 2

SWITCH 1

SWITCH 1

switch itself must be rated for the higher voltage.

Low voltage relay switching eliminates any need for line voltage switches and line voltage switch legs in the wiring system, reducing costs and adding safety. And the operating coil circuits in relay switching systems may be operated manually through the use of small pushbuttons or toggle switches or automatically through the use of photoelectric devices, time switches or other controls.

Low voltage relay switching is generally used where remote control and/or frequent individual control is required for each of a number of small 120-v or 277-v lighting loads, such as individual luminaires or small groups of closely-mounted luminaires.

In general, this type of control is confined to lighting or individual appliance loads in relatively small areas.

This would include residences, private offices, stores, restaurants, night clubs, certain rooms and small areas in schools and hospitals and other places where necessity or convenience demands multi-point and/or remote switching.

In a typical relay-controlled circuit, the relay contacts close and open the hot conductor which supplies line voltage to the one or more luminaires or other loads which the relay controls.

The relay is generally a three-wire, mechanically-held device with a distinct internal circuit for opening the contacts and another for closing the contacts, although one system uses a single-winding relay to alternately open and close contacts.

The main relay contacts will open or close depending upon which operating circuit is energized from the low voltage source. A step-down control transformer of very low capacity (a typical case might be 25 voltamperes) is generally used to provide the low voltage. And depending on the system, the transformer might be used for a single relay or for a number of relays. In any case, on-off control of the load is exercised by means of some small switching device which actually controls low voltage energy flow to the relay coil. Operation of the relay contacts is then a secondary switching action in response to the switching of the low voltage circuit.

Installation

Design and layout of low voltage relay switching systems can be made in many ways depending upon the relay components used, the branch circuit conditions and the particular control requirements.

In some cases, all of the relay units may be mounted in their own separate enclosure near the panelboard containing the branch circuits which the relays switch. This case would be typical of commercial application where each relay controls all the lighting on one branch circuit. In such a case, the transformer which supplies the low voltage could be mounted in the same enclosure. In applications where a single panelboard serves a large number of lighting branch circuits over a very large area, such as large office areas in commercial or industrial buildings, a number of relays associated with each section of the overall area may be group-mounted in an enclosure in that area. Still another layout possibility—which offers best results when individual outlets are to be controlled—involves installation of individual relays in outlet boxes.

Another type of low voltage relay switching system makes use of combination relay-transformer units. In these units, the transformer has to supply low voltage to only one relay. The compact units are made for ready installation in outlet boxes and will provide control for loads ranging from one outlet to a complete circuit.

THREE BASIC SYSTEMS

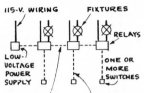

A. CENTRAL POWER SOURCE; 2-WIRE SWITCHING: one power supply serves all relays; switches do not have separate "on" and "off" contacts; pressing switch alternately turns fixture on and off.

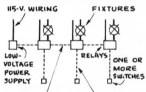

B. CENTRAL POWER SOURCE; 3-WIRE SWITCHING: One power supply serves all relays; momentary-contact switches have separate "on" and "off" positions.

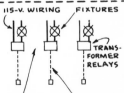

C. LOCAL POWER SOURCE; 3-WIRE SWITCHING: Transformer is built into each relay; switches have "on" and "off" positions.

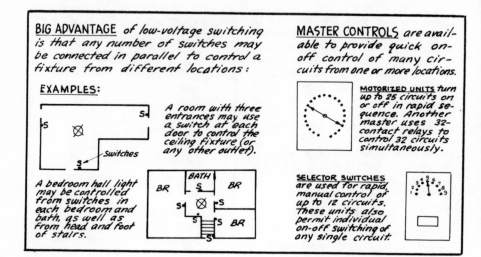

BIG ADVANTAGE of low-voltage switching is that any number of switches may be connected in parallel to control a fixture from different locations:

EXAMPLES:

A room with three entrances may use a switch at each door to control the ceiling fixture (or any other outlet).

A bedroom hall light may be controlled from switches in each bedroom and bath, as well as from head and foot of stairs.

MASTER CONTROLS are available to provide quick on-off control of many circuits from one or more locations.

MOTORIZED UNITS turn up to 25 circuits on or off in rapid sequence. Another master uses 32-contact relays to control 32 circuits simultaneously.

SELECTOR SWITCHES are used for rapid, manual control of up to 12 circuits. These units also permit individual on-off switching of any single circuit.

With this system, the transformer primary is connected to the line conductors at the point of installation in the outlet box. The connection of the relay contacts into the hot conductor is also made in the box. From the outlet box, only the low voltage control wiring has to be carried down to the low voltage wall switch.

Determination of the exact layout for any low voltage switching system will usually depend upon the size of loads to be controlled by each relay.

It invariably resolves itself into an analysis of relative lengths of line voltage conductors and low voltage conductors for various placements of relays. However, the type of system to be used in any case is also an important factor in the best layout.

The manufacturers of the many complete low voltage switching systems provide thorough, detailed literature covering all phases of design and installation of their systems.

In general, installation of all the available systems is relatively simple. All of the wiring comes under Class 2 Remote Control systems in Article 275 of the National Electrical Code. Regular bell wiring can be used for the low voltage circuit and it need not be installed in raceway. Installation of low voltage switches does not require use of switch boxes.

Time Switches

ONE OF THE FASTEST GROW-ING applications in modern electrical control is that of time switches. A time switch, as its name implies, is a control switch incorporating a mechanism which permits timed operation of its contacts.

The basic time clock contains a clock-type timing mechanism to open and close one or more sets of contacts at predetermined times. The timing mechanism is an electrically-driven clock—or may combine electric motor drive with spring drive. The unit provides simple ON-OFF control of electric circuits, operating automatically on the pre-set schedule, switching its load circuits, for as long as the switch is in use.

Construction

Typical time switches are rated up to 55-amps per pole, with one, two or three poles, single-throw or double throw, for use on circuits rated at 24, 120, 208, 240 or even 440 (with a transformer or dropping resistor) volts. Models are made with dials calibrated to provide scheduled switching over each 24-hour period or for 7-day calendar settings, with up to 288 operations per day and with a variety of dials for dividing the day into hours or minutes or for programming on the basis of sunlight or days of the week. Calendar dials designate days of the week. Astronomic dials provide control switching according to sun time for each season of the year. Other dials provide for skipping switch operations for one or more days in a weekly switching schedule.

To meet varying application requirements, time switches are available in models for regular or heavy duty. Housings are made for both indoor and outdoor applications, with sufficient knockouts to afford ready circuiting from top, bottom or sides. Models are also available with power cords and plugs for connection to convenience receptacles and with a

TIMED CONTROL of load may be accomplished directly through the substitution of a time switch in place of the conventional switch. In this case the time switch must be rated to carry the full load current.

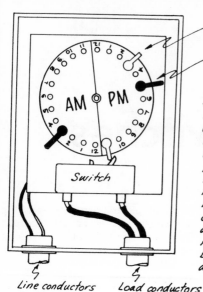

ON tripper engages switch lever, turning on switch

OFF tripper engages release lever, spring returns switch to OFF position.

AM ⊙ PM

Switch

TYPICAL TIME SWITCH

ON and OFF trippers are fastened to dial by thumb set screws at desired times of operation. Trippers shown here are placed to turn load on at 3:00 and 11:00 P.M., off at 5:00 P.M. and 3:00 A.M. Additional trippers may be used as needed for additional cycles. Mechanical configuration of trippers and switch mechanism determines minimum elapsed time permitted between successive ON and OFF operations.

Line conductors to timer motor

Load conductors to switch terminals

receptacle built into the case for connection of plug-in load appliances or devices that are mobile.

Application

Time switches offer important advantages in modern control of all types of power, lighting, heating and air conditioning equipment. Applications are almost unlimited—for both indoor and outdoor use in commercial, industrial, institutional and residential occupancies. Typical indoor applications include control of: advertising signs, store window lighting, floodlighting, electric heating, ventilation, air conditioning, refrigeration, chemical and food processes, industrial ovens, bells, signal lights, buzzers, farm operations and various types of motor loads.

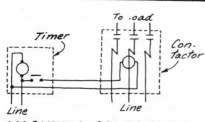

To load

Timer

Con-tactor

Line

Line

MAGNETIC CONTACTOR

may be used to control the load, the time switch being used to turn the contactor coil on and off. Time switch carries only contactor coil current.

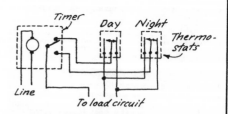

Timer

Day

Night

Thermo-stats

Line

To load circuit

MULTIPLE-CONTACT switches

in timer may be used to control two load functions, such as day-night operation of a heating system.

Typical outdoor applications include control of: street lighting, outdoor lighting circuits, parking lot lighting, irrigation and pump control and oil well pumping. The housings of such units must be rain-tight.

In use, time switches may be used to control either load circuits or control circuits. When used to control load circuits, the contacts in the switch directly open and close the conductors carrying current to the load. Time switches are available for directly switching lighting and heating circuits up to their rated current. Units are also made for directly switching motor circuits up to about 2 hp, either single-phase or three-phase.

Very wide application is also found for time switches in the ON-OFF operation of control circuits. Such would be their application as pilot control devices in combination with magnetic contactors and motor starters. In this type of control, the time switch is used to make or break the supply circuit to the operating coil of the magnetically actuated device. The main contacts of the contactor or motor starter then control the flow of current to the lighting. heating or motor load.

A single-pole, double-throw time switch must be used with mechanically-held contactors to provide a separate "make" action for the "open" and "close" circuits to the contactor coil. For use in switching the coil circuit of a magnetically-held contactor or a magnetic motor starter, a single-pole, single-throw time switch is required.

Many time switches are available with "reserve power" or "carryover". This is the inclusion of a precision spring mechanism which takes over to keep the clock mechanism operating for as long as 20 hours after power failure to the electric drive motor. This spring is wound automatically by the electric drive motor during normal operation. This feature in a time switch eliminates the need to re-set the clock dial if the power supply is restored before the end of the carryover period.

Selection

Basic selection of the proper type and size of time switch to use for any particular appilcation can be made from a series of determinations, as follows:

1. How often must the switching device be operated?—at sunset and sunrise—different times on different days of the week—every hour—or every few minutes.

2. What is the required durations of ON operation?—hours or minutes or seconds.

3. Does the application require the advantages of carryover for periods of power outage.

4. What type of dial is needed for the job?—plain 24-hour dial for the same ON-OFF switching times each day—weekend skip dial for same-time switching each day but skipping preselected days—7-day dial for different times of switching each day of the week to program switching operations for a week—astronomic dial for ON at sunset, OFF at selected time or sunrise, changing switch settings for seasonal changes in sun schedule—program dials for scheduling many ON-OFF operations in one day.

5. What type of contact arrangement is best suited to the job?—SPST for direct switching of lighting loads or for operation of operating coils in contactors or motor starters—DPST for 220-volt, 3-wire circuits or for simultaneous switching of two 120-volt circuits—3PST for switching 3-phase 3-wire or 4-wire circuits—SPDT for transferring power from one circuit to another or for mechanically-held contactors — DPDT for transfer of two-hot-wire circuits.

Relays

WITH MODERN development in art and technology of electric control, the application of electric relays has become commonplace in all types of electrical systems.

Although relays are generally very small devices, they play very important roles wherever they are used. Proper application and operation of relays is essential to effective operation of devices, circuits and systems with which they are used. A little relay can make or break the biggest power system, depending upon how it is applied in the control layout.

Construction

Basically, a relay is a magnetically-operated switching device used for a wide range of control operations. It is somewhat similar to the magnetic contactor, except that it is used for lighter duty applications and operates on lower currents. A typical relay consists of an electromagnet assembly which moves an armature to open or close one or more sets of normally-open and/or normally-closed con-

tacts. An electromagnetic coil in the unit is energized to operate the contacts. In general, relays are more often used to open and close control circuits than to operate power circuits.

As in the case of magnetic contactors, relays are made with either magnetically-held contacts or mechanically-held contacts, to meet the requirements of different applications. In the magnetically-held type, the coil of the unit must be constantly energized for as long as the contacts are to be held open or closed. In the mechanically-held type, the coil need be energized only instantaneously to open the contacts and only instantaneously to close the contacts.

Application

Various basic control actions can be performed with relays. Relays can be used to operate coil circuits in magnetic contactors, to switch the coil circuits of other relays which in turn perform a second switching operation at a higher power level, to control solenoids and to directly switch low-current motor and lighting circuits.

Relays offer a means of control circuit switching at power levels higher than the capacities of initiating pilot devices like push-buttons, float switches, pressure switches, etc. Relays with many sets of contacts can be used to convert a single control signal (to the coil) into various combinations and sequences of control switching.

A very common application for relays is overcurrent protection for motors. A relay in such a circuit is placed with its coil in the power line to the motor and is adjusted to operate its associated contacts when the value of current flowing through it reaches an unsafe level for the motor.

Still another use of relays is in interlocking of control circuits. In such cases, the coil of a relay is connected in one or more other circuits which are to be switched only after the first circuit is energized.

Selection of relays for specific applications is a relatively simple, straightforward procedure in which the specifications for the relay are related to the electrical characteristics of the circuit in which it is to be used.

For instance, the contacts on a relay must be capable of handling the maximum possible inrush current which flows when the contacts close. This inrush current will depend upon the type and size of load device which is to be supplied through the contacts.

Incandescent lamps, for instance, have inrush currents on starting which are much higher than their normal running currents This is due to the fact that the resistance of the filament is much lower when it is cold than it is when it is hot. The relay contacts must, therefore, be able to carry this high starting current. This is also true for motor starting currents supplied through relays.

Manufacturer's data should be checked on inrush current at given

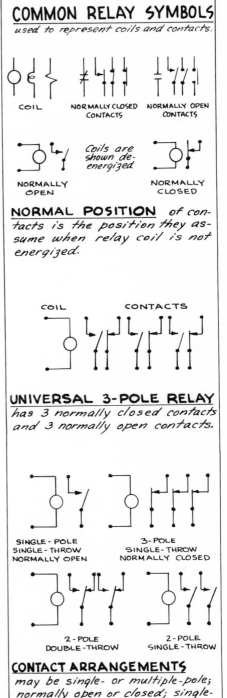

COMMON RELAY SYMBOLS
used to represent coils and contacts.

COIL   NORMALLY CLOSED CONTACTS   NORMALLY OPEN CONTACTS

NORMALLY OPEN   Coils are shown de-energized   NORMALLY CLOSED

NORMAL POSITION of contacts is the position they assume when relay coil is not energized.

COIL   CONTACTS

UNIVERSAL 3-POLE RELAY
has 3 normally closed contacts and 3 normally open contacts.

SINGLE-POLE SINGLE-THROW NORMALLY OPEN   3-POLE SINGLE-THROW NORMALLY CLOSED

2-POLE DOUBLE-THROW   2-POLE SINGLE-THROW

CONTACT ARRANGEMENTS
may be single- or multiple-pole; normally open or closed; single- or double-throw.

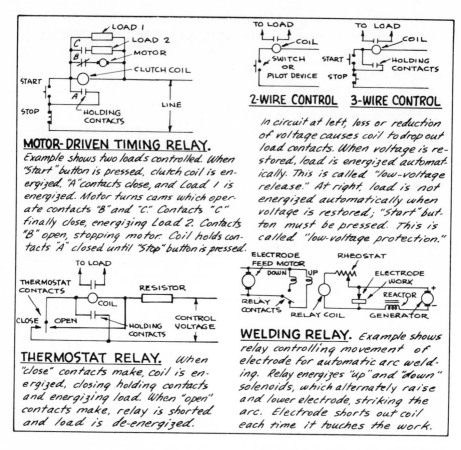

MOTOR-DRIVEN TIMING RELAY.

Example shows two loads controlled. When "Start" button is pressed, clutch coil is energized. "A" contacts close, and Load 1 is energized. Motor turns cams which operate contacts "B" and "C". Contacts "C" finally close, energizing Load 2. Contacts "B" open, stopping motor. Coil holds contacts "A" closed until "Stop" button is pressed.

2-WIRE CONTROL   3-WIRE CONTROL

In circuit at left, loss or reduction of voltage causes coil to drop out load contacts. When voltage is restored, load is energized automatically. This is called "low-voltage release." At right, load is not energized automatically when voltage is restored; "Start" button must be pressed. This is called "low-voltage protection."

THERMOSTAT RELAY.

When "close" contacts make, coil is energized, closing holding contacts and energizing load. When "open" contacts make, relay is shorted and load is de-energized.

WELDING RELAY.

Example shows relay controlling movement of electrode for automatic arc welding. Relay energizes "up" and "down" solenoids, which alternately raise and lower electrode, striking the arc. Electrode shorts out coil each time it touches the work.

voltage rating for incandescent lighting loads and other non-inductive loads and for motor loads and other inductive loads. Then the continuous current which the contacts must carry will further indicate required specs.

Finally, the maximum current which the relay will be called upon to interrupt must be determined. Again, this current value will vary with the type and size of load. And, as in the case of magnetic contactors, a relay used to directly control an ac motor load must be able to interrupt the locked-rotor current of the load.

Common control relays are rated at 10- or 25-amps per pole for switching circuits up to 600 volts. Units are made with various combinations of normally-open and normally-closed contacts, up to 12 poles. For direct control application in motor circuits, some relays are rated in horse-power. Typical voltage ratings of relay coils are: 6, 12, 24, 32, 64, 110, 220, 440, 550 and 600 volts.

A wide range of pilot devices are commonly used to control the coil circuits in relays. For 2-wire control circuits (magnetically held relays), typical pilot devices are automatic and include—single-pole, single-throw thermostats or humidistats; single-pole float or pressure switches; or interlocking or auxiliary contacts in some other relay or contactor.

For usual 3-wire control circuits, standard hookups of Start-Stop momentary-contact pushbuttons are used, with or without a holding contact circuit depending upon the type of re-

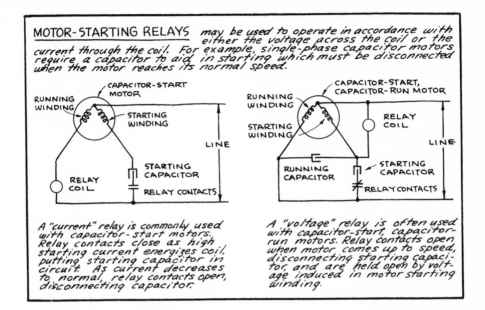

MOTOR-STARTING RELAYS may be used to operate in accordance with either the voltage across the coil or the current through the coil. For example, single-phase capacitor motors require a capacitor to aid in starting which must be disconnected when the motor reaches its normal speed.

A "current" relay is commonly used with capacitor-start motors. Relay contacts close as high starting current energizes coil, putting starting capacitor in circuit. As current decreases to normal, relay contacts open, disconnecting capacitor.

A "voltage" relay is often used with capacitor-start, capacitor-run motors. Relay contacts open when motor comes up to speed, disconnecting starting capacitor, and are held open by voltage induced in motor starting winding.

lay. For mechanically-held relays, momentary-contact pushbuttons or toggle switches can be used.

The variety of control relays in common use today is growing constantly. In addition to the conventional types of electromagnetic relays, static relay devices which contain no moving parts have already passed from the development stage to the application stage.

Other conventional relays are made with built-in step-down control transformers—primaries rated from 110 volts to 550 volts, and secondaries rated from 6 volts to 110 volts—to provide use of the relays in power circuits up to 600 volts with the control circuit to the coil operating at a lower voltage for low voltage pilot devices. The types of ac and dc relays made for specific applications include: welding relays, close differential relays for protective purposes, machine tool relays and motor operated timing relays.

CONTROL RELAYS of many different types are available with a wide variety of contact arrangements or special features making them particularly suitable for certain applications or duty requirements.

More detailed knowledge of those types more commonly used, such as general-purpose, timing, electronic, photoelectric, and low-voltage control relays, will help to insure recommending the correct relay for the job at hand.

• **General-purpose** control relays for industrial applications are used extensively for machine tool controls in the sequencing of boring, drilling, tapping, milling and grinding operations; for control of oil burners, unit heaters, electric furnaces, and stokers; and for any other operation where accurate starting, stopping, or change in motion is required. They are for the most part electromagnetic relays, with differences in contact arrangements and ruggedness, rated to handle the currents they are expected to carry or interrupt.

They are described also as having a certain number of poles, as a 2-pole relay, 4-pole relay, etc. This refers to the number of contact arrangements available. Thus a 4-pole relay would have 4 sets of contacts, prearranged

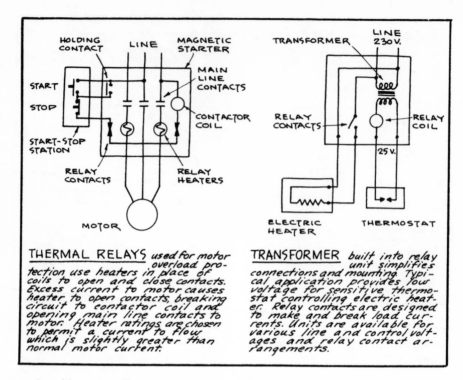

THERMAL RELAYS *used for motor overload protection use heaters in place of coils to open and close contacts. Excess current to motor causes heater to open contacts, breaking circuit to contactor coil and opening main line contacts to motor. Heater ratings are chosen to permit a current to flow which is slightly greater than normal motor current.*

TRANSFORMER *built into relay unit simplifies connections and mounting. Typical application provides low voltage for sensitive thermostat controlling electric heater. Relay contacts are designed to make and break load currents. Units are available for various line and control voltages and relay contact arrangements.*

to be either normally open, normally closed, or a combination of the two.

● **Universal** relays are built with a variety of contact arrangements which permit wiring the contacts as either normally open or normally closed, without replacing or reassembling any parts.

A 4-pole universal relay would have a set of normally closed and a set of normally open contacts for each of the 4 poles. Thus a fewer number of relay types need be stocked for required applications.

● **Timing** relays of many types are available to provide the time delay often desired between the closing of a control circuit and the actual energizing of a load.

A typical timing relay may control two loads—one instantaneously, one after a short delay. For example, the instantaneous load could be a solenoid valve which admits lubrication to a pump; the pump motor then starts

after a short delay after lubrication has been accomplished.

Another application might be to keep a fan motor running for several minutes to dissipate heat from the plenum of a warm-air furnace after the actual burning of fuel has been terminated.

A short time delay is often obtained by designing the electromagnetic relay with a copper jacket over the iron core. Induced current in the copper opposes the main magnetic field tending to close the contacts, introducing a delay in operation.

Delays in de-energizing a relay may also be obtained through use of an electrolytic capacitor connected across the relay. The capacitor stores energy while the coil is energized.

When the current to the coil is cut off, the capacitor discharges its stored energy through the coil, maintaining the relay contacts in their closed position for a short time which is determined by the amount of capacitance and resistance in the circuit.

• **Motor-driven Timing** relays provide accurate timing periods of longer duration due to the flexibility possible in the design of the motor and the mechanical linkage controlling the relay contacts.

Timing periods available from a single unit may have a range of from several seconds to six or more hours. The relay consists of a synchronous motor, any desired combination of contacts to switch the load, and a solenoid-operated clutch. When the pushbutton or other activating device energizes the relay, the clutch engages a gear train which is driven by the motor. The gears cause a cam to rotate until the relay contacts are opened (or closed). The adjustment of the cam determines the time required to operate the contacts. After the motor is disconnected or stopped, the magnetic action of the solenoid holds the relay contacts open (or closed) until the activating device initiates another cycle.

• **Electronic Timing** relays use an electron tube with a suitable resistor-capacitor circuit. The timing is accomplished by controlling the passage of current through the electron tube through the charging or discharging of the capacitor.

• **Thermostat** type relays are used with low-voltage, 3-wire thermostats, making it unnecessary for the thermostat to make and break the circuit to the heating or cooling device.

The relay closes when the thermostat "common" contact makes with a "close" contact; the relay opens when the common contact makes with an "open" contact. Thus the "close" contact may cause the relay to connect an electric heater, which then remains on until the "open" contact causes the relay to disconnect the heater.

• **Photoelectric** relays are activated either by the presence or absence of light, such as for combustion controls, smoke detection, street lighting, or intruder alarms. A typical unit contains a photocell, an electron tube, and an electromagnetic relay. When light hits the photocell, it changes the voltage on the grid of the electron tube. The grid acts as a valve, permitting current to pass through it and to the relay coil, thus closing the relay contacts. By revising the wiring to the electron tube and the photocell, the same unit may be used to close the relay contacts when the light goes out or is interrupted.

• **Low-voltage** control relays used for remote switching of lighting and appliance circuits are of three general types. One consists of a single coil which alternately opens or closes its contacts upon momentary current impulses from a pushbutton. Thus, pressing the button turns on the light or other load; pressing it again turns off the light.

The second type uses two coils. A momentary current provided by pressing an "on" switch energizes one of the coils, turning on the load. Pressing the "off" switch energizes the other coil and turns off the load.

The third type is a thermal relay having no coil. Current from the "on" switch heats a sensitive "on" thermal wire, causing the contacts to close by expansion of the wire.

A second thermal wire acts similarly to open the contacts. All three types are single-pole, mechanically latching relays. The thermal type mentioned has its own built-in transformer to provide the low operating voltage; the others depend upon an external transformer.

Magnetic Contactors

MAGNETIC contactors are switching devices in which the opening and closing of the switch contacts is actuated by electromagnetic action. The electromagnetic action is derived from passing current through a coil, making it an electromagnet which exerts a force to move one set of contacts against a stationary set of contacts.

In general operation, magnetic contactors are similar to magnetic motor starters. The only difference between the types of devices is that the motor starter contains overload protective devices (usually thermal relay units) which are required to protect the motor.

The basic type of magnetic contactor is operated by energizing the magnetic operating coil to close the contacts and is magnetically held in the closed position by maintaining current flow through the coil.

If the voltage supply to the coil is removed, either intentionally or through fault in the circuit, the contacts will open. The coil must be constantly energized to keep the contacts in the closed position. All the time the contactor is closed, therefore, the coil is consuming electric energy.

Magnetically held contactors are finding ever wider application in modern power and light systems for remote and/or automatic switching of ac and dc loads. This type of switch control of loads is of particular advantage where frequent opening and closing of the circuit is a requirement.

Standard contactor units are made and rated for high inrush current loads (tungsten filament—incandescent—lamps), other resistance loads, fluorescent lighting loads and to some extent for motor loads.

A switching device used to control motors must be capable of interrupting locked-rotor (or stalled-rotor) current of the motor loads. This is the current drawn by a motor when voltage is applied to the stator while the rotor is not rotating. It is the condition of maximum operating overload. Locked-rotor current may run anywhere from 3 to 8 times the normal full-load running current of the motor.

A contactor may be used to control a motor load up to the limit of its rated ability to interrupt locked-rotor current. Of course, when a contactor is used to control a motor, some provision must be made to provide required protection against running overload. This protection is normally included in the enclosure of standard motor starters.

To facilitate selection of the proper pilot devices for control circuits of magnetic contactors, some catalogs and bulletins indicate operating coil currents and wattage ratings of coils in different sizes of contactors.

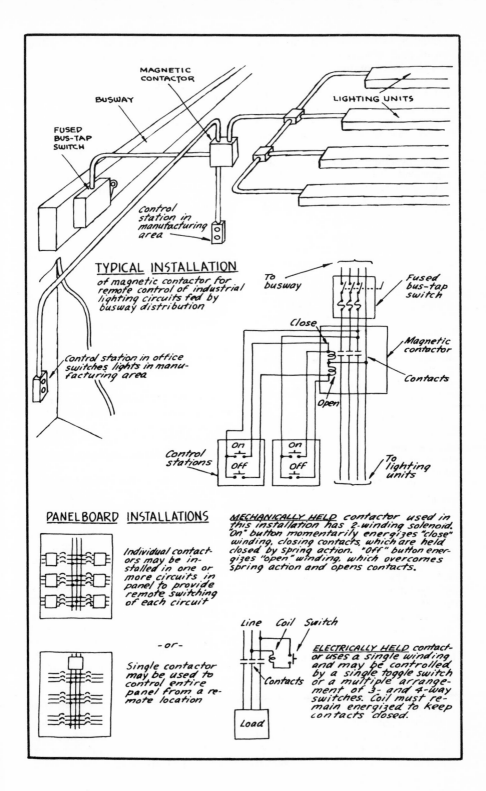

MAGNETIC
CONTACTOR

BUSWAY

FUSED
BUS-TAP
SWITCH

LIGHTING UNITS

Control
station in
manufacturing
area

TYPICAL INSTALLATION
of magnetic contactor for
remote control of industrial
lighting circuits fed by
busway distribution

To
busway

Fused
bus-tap
switch

Close

Magnetic
contactor

Contacts

Open

Control station in office
switches lights in manu-
facturing area

Control
stations

On
OFF

On
OFF

To
lighting
units

PANELBOARD INSTALLATIONS

Individual contact-
ors may be in-
stalled in one or
more circuits in
panel to provide
remote switching
of each circuit

-or-

Single contactor
may be used to
control entire
panel from a re-
mote location

<u>MECHANICALLY HELD</u> contactor used in
this installation has 2-winding solenoid.
"On" button momentarily energizes "close"
winding, closing contacts, which are held
closed by spring action. "Off" button energizes "open" winding, which overcomes
spring action and opens contacts.

Line   Coil   Switch

Contacts

Load

<u>ELECTRICALLY HELD</u> contact-
or uses a single winding
and may be controlled
by a single toggle switch
or a multiple arrange-
ment of 3- and 4-way
switches. Coil must re-
main energized to keep
contacts closed.

Another factor involved in selection of pilot devices is the voltage of the coil circuit of the contactor. The operation coil in a contactor may be operated at the same voltage as that of the power or light circuit which the contactor switches or at some lower voltage obtained through a control transformer. And if momentary-contact pushbuttons are used with a magnetically held contactor, a holding circuit must be established through an auxiliary set of contacts in the contactor.

Such an arrangement requires three wires for the control circuit of the operating coil. If a maintained-contact device is used to energize the operating coil, only two wires are necessary for the control circuit.

For any particular application, a magnetically held contactor must be selected on the basis of the load conditions. Catalog data should be studied to determine that the ratings given for any contactor are suited to the type of load to be handled.

The mechanically held magnetic contactor varies from the basic type of magnetically held contactor in the operation of the electromagnet assembly. In the mechanically held contactor, the operating coil need be energized only momentarily to close the main contacts. Then to open the contacts, the operating coil is again momentarily energized over a separate control circuit.

During the time that the main contacts are closed or open, the contactor operating coil is drawing no current and is not energized. In the open or closed position, the contacts are maintained in their position by either some type of mechanical latch device or by a permanent magnet assembly.

Because of the definite switching action—with mechanically maintained open and closed positions—this type of magnet contactor is commonly distinguished from the magnetically held type by calling it a "remote-control switch," i.e., a mechanical switch which can be operated in remote places by means of a long "Handle"— the control circuit between the contactor and the pilot switching device being the "handle."

Operating characteristics of the mechanically held contactor give it particular advantages. Because it will not trip out (open) on voltage dips or failures—as magnetically held contactors will do—the mechanically held contactor is well suited to use on circuits for lighting or other loads where voltage fluctuations do not hurt the load devices.

In such circuits, magnetically held contactors would keep dropping out and disconnecting the load devices. Of course, if magnetically held contactors are used on circuits with good voltage regulation, they are just as acceptable as the mechanically held type.

Mechanically held contactors are currently finding widespread application in switching of power and lighting feeders and branch circuits, offering remote control from one or more control points. Such contactors are also being used to provide overall control of branch circuits in a panelboard by switching the main buses in the panel.

In other cases, split-bus panelboards have contactors built into them between the bus sections, with the contactor switching the feed to one section of the bus. In these various applications, typical units might be rated up to 100-amps, with four poles.

Magnetic contactors are generally used as remote switches with manual pushbuttons or toggle switches used to operate the control circuits. There are many applications, however, in which the control circuit to the operating coil of the contactor is operated by an automatic pilot device.

Outdoor lighting circuits and sign circuits, for instance, are commonly switched by magnetic contactors which are controlled by time switches. In still other cases, various types of relay control are used.

Motor Controls

ELECTRIC motor controls include a wide range of devices and groups of devices used to control the operation of motors. Depending upon the type, size and application of a motor to be controlled, a motor controller may be simple or complex, either manually or automatically operated. Simple motor controllers provide only starting and stopping of motors; others may provide limiting of starting current, speed control, reversal of rotation, protection against overload, protection against undervoltage. Although the basic operating principles of all motor controls are similar, motor control equipment can be divided into two groups: direct-current and alternating-current motor controls.

DC Motor Controls

Direct-current motors which draw very little current from their supply circuit when starting may be started by placing full voltage across the motor terminals. A motor controller which connects a motor to its supply circuit in such a way that full line voltage is impressed on the motor is called an *"across-the-line" starter*. This type of starting is usually confined to dc motors rated less than 2 horsepower. Across-the-line dc starters are either manually operated switches (toggle switches) or magnetic coil-operated contactors.

Manual across-the-line switches for control of dc motors may be single-throw type, providing "on" and "off" positions for simple starting and stopping the motor, or double-throw type, providing an "off" position for stopping the motor and two "on" positions for starting the motor with either clockwise or counterclockwise rotation, The double-throw type is, therefore, a reversing starter. Some manual starting switches are equipped with a thermostatic element which opens the switch or a fuse which breaks the circuit when the motor is overloaded.

Across-the-line magnetic contactors for starting dc motors are made for either simple starting and stopping or for starting, stopping and reversing. The reversing type of switch consists of two of the start-stop contactors properly wired for motor rotation in either direction. Magnetic contactors employ a solenoid

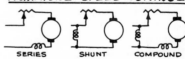

ACROSS-THE-LINE MAGNETIC SWITCH

Hold contact Armature
Line Coil
Start Stop
Main
Contacts
Shunt
Field

• Pressing START button sends current through coil. Attraction of coil closes main contacts and hold contact, starting motor. Hold contact keeps coil energized after button is released.

• STOP button cuts off current to coil, opening contacts and stopping motor.

ARMATURE SPEED CONTROL

SERIES   SHUNT   COMPOUND

The speed of all three common types of direct current motors may be controlled by varying resistance in series with the armature.

193

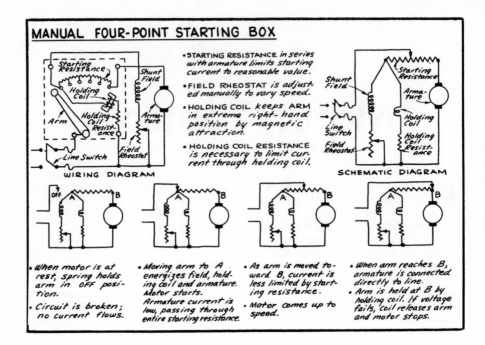

MANUAL FOUR-POINT STARTING BOX

- **STARTING RESISTANCE** *in series with armature limits starting current to reasonable value.*
- **FIELD RHEOSTAT** *is adjusted manually to vary speed.*
- **HOLDING COIL** *keeps ARM in extreme right-hand position by magnetic attraction.*
- **HOLDING COIL RESISTANCE** *is necessary to limit current through holding coil.*

WIRING DIAGRAM

SCHEMATIC DIAGRAM

- *When motor is at rest, spring holds arm in OFF position.*
- *Circuit is broken; no current flows.*

- *Moving arm to A energizes field, holding coil and armature. Motor starts. Armature current is low, passing through entire starting resistance.*

- *As arm is moved toward B, current is less limited by starting resistance.*
- *Motor comes up to speed.*

- *When arm reaches B, armature is connected directly to line.*
- *Arm is held at B by holding coil. If voltage fails, coil releases arm and motor stops.*

(or two in reversing starters) which makes and breaks the contacts in the unit. The coil in a magnetic contactor may be energized by a pushbutton, a float switch, a thermostat or some other 2-wire pilot device. Magnetic contactor units are equipped with a built-in relay which opens the circuit to the motor when an overload occurs. This relay can be reset by simply pushing a "Reset" button on the contactor.

When large dc motors are started, they draw very high current because they offer very little resistance to current flow when they are at rest. If full line voltage were applied to a large dc motor while the armature is at rest, the heavy flow of current through the motor would damage the motor or burn out the fuse. To avoid this result, large dc motors are controlled by reduced-voltage starters. In starting a large motor, a resistance unit is placed in series with the motor when it is connected across the supply line. The resistance limits the starting current to a safe value to prevent damage to the motor. Then as the motor starts

to rotate and come up to speed, the internal resistance of the motor increases, allowing gradual reduction of the amount of external resistance in series with the motor. When the motor has come to full speed, external resistance can be cut out.

Reduced-voltage motor starters are arranged in compact enclosures, including the external resistance unit and the means for gradually cutting the resistance out of the circuit. Such starters are made in both manual and automatic types.

Manual reduced-voltage starters are mounted in boxes and are commonly called "starting boxes." There are several types of starting boxes as follows:

- **Three-Point Starting Box** — This starter can be used with either a shunt or compound dc motor. The unit contains a tapped resistance element which is used for limiting starting current. When the handle on the unit is moved from resistance tap to resistance tap, the resistance in series with the motor is gradually decreased. When the handle has moved to the last resistance tap, all of the resistance has been cut out of the circuit; and a magnetic coil holds the handle in this position. The starter gets its name from the fact that there are three terminals or "points" on the face plate. One of these terminals connects to one of the supply line wires and is marked "L." Another terminal connects to the motor field and is marked "F." The third terminal connects to the motor armature and is marked "A." In use, if the shunt field should open for any reason, the magnetic coil holding the handle in the running position would be de-energized, the handle would return to its off position, removing voltage from the motor armature. This provision is called "no field release."

• **Four-Point Starting Box** — This is the most common type of starting box in general use today, and is similar to the three-point box. It has four terminals on its face plate—one for the armature, marked "A"; one for the field, marked "F"; and one for each of the line wires, marked "L_1" and "L_2." As in the case of the three-point box, this unit provides current limiting resistance for starting and the means for removing the resistance. In addition, however, the four-point box may incorporate another resistance which allows speed control of the motor as well as starting control. The four-point controller has "under-voltage"

release, which returns the controller handle to its "off" position when voltage falls below a certain value.

Manual reduced-voltage controllers are used on motors up to about 50 horsepower.

An **automatic reduced-voltage** controller is designed to control automatically the acceleration of a motor. They are usually of the pushbutton type and contain all of the devices necessary for automatically changing the resistance in series with a motor as the motor starts and comes up to speed. The functions of reduced-voltage starting are accomplished internally by electromagnets. The start-stop buttons may be remotely located from the motor. The motor is started by simply pushing the "start" button and stopped by pushing the "stop" button. Automatic starters are made for shunt, series and compound motors—for constant speed and adjustable speed motors.

A **drum controller** is another common type of controller used for starting shunt, series and compound motors. It consists of an enclosed "drum" switch and an auxiliary starting resistance. Used with a series motor, a drum controller will provide series resistance for starting and may provide for reversal of motor rotation. With series motors, drum controllers can also provide speed control by varying the resistance.

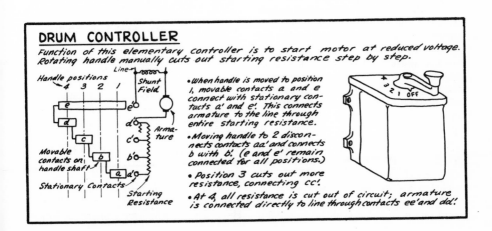

DRUM CONTROLLER

Function of this elementary controller is to start motor at reduced voltage. Rotating handle manually cuts out starting resistance step by step.

• When handle is moved to position 1, movable contacts a and e connect with stationary contacts a' and e'. This connects armature to the line through entire starting resistance.

• Moving handle to 2 disconnects contacts aa' and connects b with b'. (e and e' remain connected for all positions.)

• Position 3 cuts out more resistance, connecting cc'.

• At 4, all resistance is cut out of circuit; armature is connected directly to line through contacts ee' and dd'.

ACROSS-THE-LINE STARTER

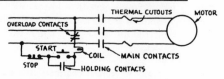

START-STOP STATION

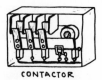

CONTACTOR

- *Pressing START button energizes coil, which closes main contacts and holding contacts, starting motor.*
- *HOLDING CONTACTS keep coil energized after start button is released by short-circuiting button.*

- *STOP button cuts off current to coil. Main contacts open, stopping motor.*
- *OVERLOAD CONTACTS protect coil, opening circuit if current becomes excessive.*
- *THERMAL CUTOUTS open motor circuit if line current becomes too great.*

CONTROLS for alternating-current motors vary widely depending upon motor application and operating characteristics. As in the case of dc motor controls, alternating-current motor controls can be divided into across-the-line types and reduced-voltage types. Controls may be simple start-stop units or may also include other functions such as speed control, rotation reversing and various types of protection—against overheating, overloading, overvoltage.

If full line voltage is connected directly to a motor, the current drawn from the line will be several times the normal running current of the motor. In the case of many large motors, the heavy current drawn during starting at full line voltage would damage the motor windings and cause voltage disturbances in the supply line. For this reason, across-the-line starting is used with small motors or where the load can stand the current inrush and where line disturbances are slight. With many motors, a reduced-voltage control unit is used to gradually apply full voltage to the motor and thereby avoid the sudden high current which accompanies full voltage starting.

- **Manual Starting Switches** are the simplest types of ac control devices. These are across-the-line type switches and include ordinary snap-action toggle switches and pushbutton switches.

Toggle-type starting switches are made with one or two poles for ac motors up to 1 horsepower, in the range of 110 and 220 volts ac. They usually include overload protection for the motor and are available with a pilot light in the same enclosure. They are used with single-phase motors in heaters, compressors, fans, pumps and machine tools.

- **Pushbutton manual across-the-line** starting switches contain electrical contact assemblies which are opened and closed by pushbuttons, providing start and stop. The pushbuttons are mechanically coupled to the contact assembly and protrude through the cover of the switch enclosure. These switches are made for use with single-, two- or three-phase motors, up to 600 volts and about 7½ horsepower. They contain overload protection for the motor and are available in arrangements which provide for reversal of motor rotation.

- **Magnetic Across-the-Line Switches** differ from manual across-the-line switches in that the contacts are opened and closed by the action of a magnetic coil. Pushbuttons or other pilot devices may be used to energize the coil and thereby control the opening and closing of the contacts. An advantage of this type of switch is remote operation—the pushbuttons may be located some distance from the switch and the motor, affording convenience and safety.

Magnetic switches are made for single-, two- or three-phase motors, from about 2 to 600 hp., for voltages of 110,

208, 220, 440 and 600 volts. They are available in a wide range of enclosures— general purpose, water-tight, dust-tight, hazardous location or corrosion-proof— and are designed for across-the-line starting of polyphase squirrel-cage motors and single-phase motors and for control of slip-ring motors. They may be used with pushbuttons, float switches, pressure switches, thermostats or other pilot devices which provide automatic and/or remote operation of motors.

Magnetic across-the-line switches are also available for reversing of polyphase squirrel-cage and slip-ring motors, in sizes up to 750 hp.

Combination magnetic across-the-line switches combine a fused or unfused disconnect switch or circuit breaker in the same enclosure with the magnetic switch. The combination switch not only provides motor control but meets the electrical code requirement for a motor disconnecting device in the supply line.

Automatic multi-speed types of magnetic across-the-line starting switches are used to control the speed of multi-speed squirrel-cage motors, such as those used on machine tools and textile machinery.

• **Reduced-Voltage Motor Controls** are used to slowly and smoothly accelerate motors up to full speed and to limit the starting current to a value which will be safe for the motor and will cause no objectionable line voltage disturbances. Such controls must apply less than line voltage to the motor to start and then gradually increase the voltage applied to the motor, until the motor is operating on full line voltage. There are several types of reduced-voltage starters:

Primary-resistance starters contain resistance units which are connected in series with the motor during starting. With less than full voltage applied to the motor, it will start to rotate slowly. As it rotates, the internal resistance of the motor itself increases, reducing the line current and allowing removal of the series resistance in the starter. The series resistance is removed gradually as the motor resistance increases gradually. When the motor has come up to full speed, the resistance of the motor is high enough to allow removal of all of the series resistance.

Primary-resistance starters are used in the stator circuit of the motor. They may be manual, in which resistance is re-

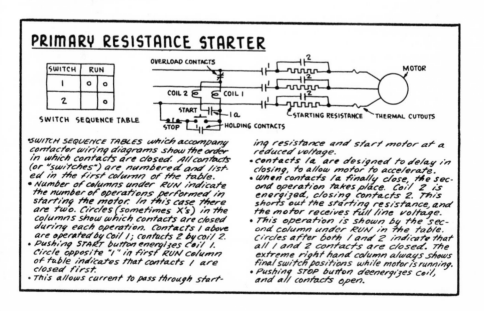

PRIMARY RESISTANCE STARTER

SWITCH	RUN	
1	o	o
2		o

SWITCH SEQUENCE TABLE

OVERLOAD CONTACTS
COIL 2 COIL 1
START
STOP
HOLDING CONTACTS
STARTING RESISTANCE   THERMAL CUTOUTS
MOTOR

• SWITCH SEQUENCE TABLES which accompany contactor wiring diagrams show the order in which contacts are closed. All contacts (or "switches") are numbered and listed in the first column of the table.
• Number of columns under RUN indicate the number of operations performed in starting the motor. In this case there are two. Circles (sometimes X's) in the columns show which contacts are closed during each operation. Contacts 1 above are operated by Coil 1; contacts 2 by coil 2.
• Pushing START button energizes Coil 1. Circle opposite "1" in first RUN column of table indicates that contacts 1 are closed first.
• This allows current to pass through start-

ing resistance and start motor at a reduced voltage.
• contacts 1a are designed to delay in closing, to allow motor to accelerate.
• when contacts 1a finally close, the second operation takes place. Coil 2 is energized, closing contacts 2. This shorts out the starting resistance, and the motor receives full line voltage.
• This operation is shown by the second column under RUN in the table. Circles after both 1 and 2 indicate that all 1 and 2 contacts are closed. The extreme right hand column always shows final switch positions while motor is running.
• Pushing STOP button deenergizes coil, and all contacts open.

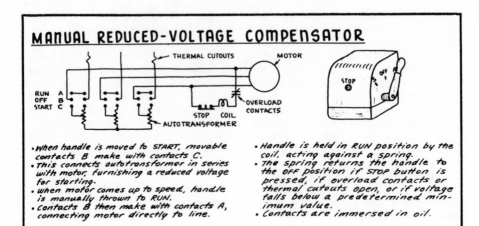

MANUAL REDUCED-VOLTAGE COMPENSATOR

THERMAL CUTOUTS
MOTOR

RUN A
OFF B
START C

STOP COIL
AUTOTRANSFORMER

OVERLOAD
CONTACTS

STOP
OFF

- When handle is moved to START, movable contacts B make with contacts C.
- This connects autotransformer in series with motor, furnishing a reduced voltage for starting.
- When motor comes up to speed, handle is manually thrown to RUN.
- Contacts B then make with contacts A, connecting motor directly to line.

- Handle is held in RUN position by the coil, acting against a spring.
- The spring returns the handle to the OFF position if STOP button is pressed, if overload contacts or thermal cutouts open, or if voltage falls below a predetermined minimum value.
- Contacts are immersed in oil.

duced by moving a handle, or they may be automatic, in which timing mechanisms reduce the series resistance in steps. They are used for starting polyphase squirrel-cage motors, 5 to about 50 hp., up to 600 volts.

A **secondary-resistance starter** is a resistance starter used in the rotor (secondary) circuit of 3-phase wound-rotor induction motors. Better control of torque is the advantage of this type of hookup. In such cases, an across-the-line magnetic starter is connected to the stator circuit and fed from the supply line. By inserting resistance in the rotor circuit during starting, line current is limited to a low value. The resistance is gradually removed as the motor comes up to speed. Wound-rotor resistance starters are made with either manual or magnetic-automatic control of resistance.

Autotransformer starters or compensators employ transformer action to provide reduced voltage to polyphase squirrel-cage motors. They have an advantage in that they do not waste energy in the form of heat, as resistance starters do. These starters are connected in the stator circuit. They provide reduced voltage and variation of voltage up to full voltage on the motor by varying the ratio of transformation, either manually with a handle or automatically by means of magnetic contactors and timing devices.

Another type of ac motor control is the **manual drum controller** for starting or reversing small 3-phase motors, for split-phase, capacitor or two-phase motors. Drum controllers are used where the motor is close to the operator—small lathes and other machine tools. Simply by moving the handle from one position to another, the motor is reversed.

Wiring Devices

THE TERM "wiring devices" is commonly used to describe the very wide range of types and sizes of convenience receptacles, cord plugs, wiring switches for control of branch circuits and individual lamps or appliances, outlet boxes, utility boxes, switch boxes and lamp sockets and holders. These small devices are used on all types of branch circuits to provide both permanent and separable connections for utilizing available power, to protect the circuit conductors, to provide control and to afford maximum protection to the user of the available power.

Receptacles

Convenience or plug receptacles are made in many types and sizes to provide cord connection of lighting units, motor operated devices and other appliances to branch circuits. Receptacles afford ready compliance with National Electrical Code requirements on ratings and applications of separable connections.

A typical convenience receptacle consists of two or more stationary contacts mounted in a small insulating enclosure, with slots to provide entry of blades on attachment plugs to make contact with the circuit. **A typical attachment plug or cap** consists of two or more projecting blades set in a small insulating base, with provision for connecting the plug to a cord.

Receptacles are generally classified according to the number and arrangement of the contact slots and according to the type of mounting. Receptacles are made with the following slot constructions: two slots, for parallel blades; two slots, for tandem blades; two slots, for polarized blades; and four slots, for four blades. According to mounting, there are flush-mounting receptacles, surface-mounting receptacles and receptacles for use in special enclosures. Flush-mounting receptacles are made in three lines: a standard line, an interchangeable line and a combination line.

The standard line of receptacles is made larger and more ruggedly than the other two lines, and is used for relatively heavy duty applications. These receptacles are made with supporting straps and two sets of mounting screws. They are available in single and duplex types, with two, three and four poles, up to 20 amps, 250 volts. Regular two-pole units with parallel slots will accommodate only parallel blades.

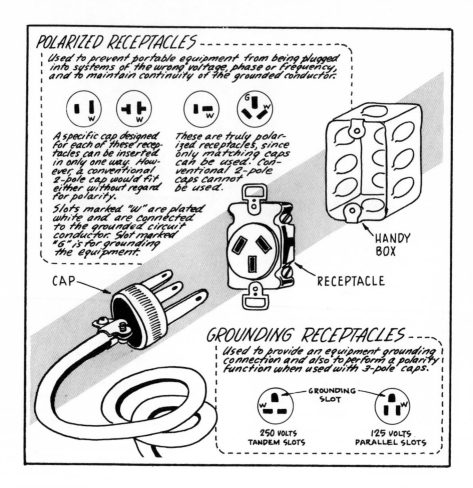

POLARIZED RECEPTACLES

Used to prevent portable equipment from being plugged into systems of the wrong voltage, phase or frequency, and to maintain continuity of the grounded conductor.

A specific cap designed for each of these receptacles can be inserted in only one way. However, a conventional 2-pole cap would fit either without regard for polarity.

These are truly polarized receptacles, since only matching caps can be used. Conventional 2-pole caps cannot be used.

Slots marked "W" are plated white and are connected to the grounded circuit conductor. Slot marked "G" is for grounding the equipment.

CAP

HANDY BOX

RECEPTACLE

GROUNDING RECEPTACLES

Used to provide an equipment grounding connection and also to perform a polarity function when used with 3-pole caps.

GROUNDING SLOT

250 VOLTS
TANDEM SLOTS

125 VOLTS
PARALLEL SLOTS

Units with "T" slots will take parallel or tandem blades.

Polarized receptacles are used with matching plugs to connect appliances to specific supply conductors, such as three-wire, single-phase or four-wire, three-phase or dc—where polarization of contacts is necessary to prevent improper or unsafe connection. These are made up to 60-amps, 250 volts.

Grounding type receptacles used for general purpose convenience outlets have two current carrying contacts and one grounding contact. The grounding contact is electrically connected to the yoke and also to a terminal on the side of the unit. A grounding conductor used to provide an equipment ground in non-metallic raceway systems is connected to the grounding terminal on such receptacles. In metallic raceway systems, equipment grounding is made through the metallic raceway and enclosures. In such cases, the yoke is connected to the metallic system by the mounting screws, providing the necessary ground for the grounding contact. These receptacles readily accommodate standard two-pole, parallel-blade plugs.

Another commonly used type of receptacle is the **twist-type.** In these units, the plug is attached by a pushing and turning action which locks the plug in the connected position. Receptacles of the twist-lock type take matching twist-lock plugs. Receptacles

WEATHERPROOF RECEPTACLE

For outdoor use or indoors where excessive moisture is present.

RANGE OUTLET

Heavy-duty receptacle made especially for electric ranges also finds use in industrial installations.

CLOCK HANGER RECEPTACLE

Recessed space for cord storage allows clock to hang flat.

of the "twist-tite" type take standard parallel bladed caps. Twist type receptacles are made in a variety of single and duplex models for two, three and four blades.

Although household electric appliances are generally connected to their supply circuits by the use of the above described types of receptacles, the electric range is commonly connected by a special range receptacle and connector. Such units are made in flush and surface types.

The interchangeable line of receptacles consists of much more compact units, smaller in size and rating than the standard line. As many as three such interchangeable receptacle units may be grouped in one gang space in an outlet box. The interchangeable devices are not suited to heavy-duty applications. Only two-pole receptacles are made in this type, rated at 15-amps for 125 volts, 10-amps at 250 volts. The interchangeable line gets its name from the fact that different combinations of interchangeable size switches, receptacles and pilot lights, up to three units, may be assembled on a single supporting strap yoke and mounted in a single gang space.

Flush type receptacles are made in many more specific types than those described here. All flush receptacles

are designed for mounting inside standard sectionalized switch boxes, outlet boxes or conduit fittings. The receptacle in the box is then covered with an outlet box cover, a conduit fitting cover or a standard flush plate. There are a variety of ways in which receptacles may be mounted in boxes.

The combination line of receptacles is very much like the interchangeable line and finds the same light-duty applications. Combination devices consist of various combinations of three units—receptacles, switches, and/or pilot lights—mounted in one molded composition body. A combination device fits in a single gang box.

Surface mounting receptacles are made in units of one, two and three receptacles—for ordinary two-pole plugs, for various light and heavy-duty polarized, three- and four-wire plugs.

Heavy-duty receptacles mounted as part of special conduit fittings are used for many industrial plug-connection applications requiring high current at higher utilization voltages. Typical of such receptacles are watertight, weatherproof and hazardous-location units—with their covers. Such units are made in wide ranges of sizes for all types of residential, commercial and industrial applications.

Special receptacles with recessed

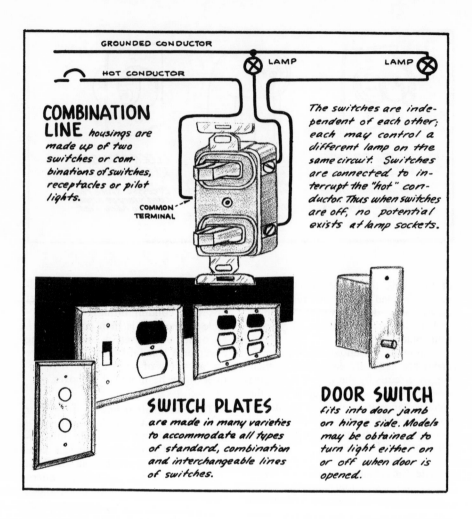

COMBINATION LINE housings are made up of two switches or combinations of switches, receptacles or pilot lights.

COMMON TERMINAL

The switches are independent of each other; each may control a different lamp on the same circuit. Switches are connected to interrupt the "hot" conductor. Thus when switches are off, no potential exists at lamp sockets.

GROUNDED CONDUCTOR

HOT CONDUCTOR

LAMP   LAMP

SWITCH PLATES are made in many varieties to accommodate all types of standard, combination and interchangeable lines of switches.

DOOR SWITCH fits into door jamb on hinge side. Models may be obtained to turn light either on or off when door is opened.

fronts, designed for flush mounting, are made for attaching cords from electric clocks and fans, with provision for supporting the clock or fan on the wall surface directly over the receptacle units.

A WIDE range of relatively small switches is included under the term "wiring devices." Such wiring switches are used in just about every type of interior wiring system to control branch circuit equipment—lighting units or appliances.

Generally, these switches operate on a snap-action principle, in which

either a toggle handle, pushbuttons or a rotary handle are used to operate a spring-and-cam mechanism (either rotating or rocking motion) which gives definite "on" and "off" positions.

Flush types of wiring switches usually consist of a basic switching mechanism set in a plastic, porcelain or composition molded case, with steel frame members for mounting the unit in switch or outlet boxes.

Standard Switches

The standard line of flush switches covers the more rugged types which are made sturdy for use in installations where the switch will be operated often to carry heavy currents. **Standard flush tumbler** switches, with toggle handles, are made in both spring-and-cam types and mercury types. Typical ratings for standard switches are as follows: 10-amp, 125-volt; 5-amp, 250-volt; 20-amp, 125-volt or 250-volt; 30-amp, 125-volt or 250-volt. These switches include single-pole and double-pole units, three-way and four-way switches and top-wired and side-wired types. Special units are made for shallow mounting applications.

Mercury switches consist of a small tubular glass envelope in which a small amount of mercury is sealed. Two contact wires are brought into the envelope at one end, positioned close together. The switch handle provides rocking motion of the glass tube, causing the liquid mercury to roll to either end of the tube depending upon the handle position. When the mercury is at the end containing the two contacts, the mercury provides a conducting path between the contacts—which is the "on" position of the switch.

Mercury switches are almost completely silent in operation, with no click or other snapping noise when the handle is moved to either position. Such switches find use where silent operation is advantageous—hospitals, nurseries, etc.

Of course, the high quality and long-life reliability of these switches have made them popular for general use in quality wiring installations—residential and commercial occupancies, particularly—where silent operation is not essential.

Although the toggle-handle tumbler type switch with definite "on" and "off" positions is the most commonly used wiring switch, there are other types for special applications. The standard toggle-handle switch is available with **momentary contact** operation, and the standard "on-off" mechanism is available with lock-type operation.

Locking-type switches do not have protruding handles but are operated by a key inserted in a slot in the face of the switch cover. Such switches are used where it is required to assure only authorized operation of switches.

As described in the National Electrical Code, snap switches are given three ratings as follows:

1. **Non-Inductive loads:** for non-inductive loads (loads which contain no transformers, motors or other electro-magnetic windings) other than tungsten-filament lamps, switches must have an ampere rating not less than the ampere rating of the load.

2. **Tungsten filament loads:** For tungsten-filament lamp loads and for combined tungsten-filament and non-inductive loads, switches must be "T" rated, except where the three following qualifications are satisfied:

a. If switches are used in branch circuit wiring systems in private homes, in rooms in multiple-occupancy dwellings used only as living quarters, in private hospital or hotel rooms, or in similar places but not in public rooms or places of assembly; and

b. Only when such a switch controls permanently connected fixtures or lighting outlets in one room only, or in one continuous hallway, or in attics or basements not used for assembly; and

c. When the switch is rated at not less than 10-amp, 125-volt; 5-amp, 250-volt and for the four-way types, 5-amp, 125-volt; 2-amp, 250-volt.

3. **Inductive loads:** Switches controlling inductive loads must have an ampere rating twice the ampere rating of the load, unless they are approved for a particular use.

Catalog data on wiring switches generally indicates **"T" ratings** of switches. Although it might seem a rather involved problem to select switches on the basis of the above re-

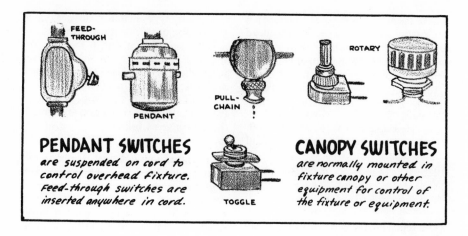

PENDANT SWITCHES
are suspended on cord to control overhead fixture. Feed-through switches are inserted anywhere in cord.

FEED-THROUGH

PENDANT

PULL-CHAIN

TOGGLE

ROTARY

CANOPY SWITCHES
are normally mounted in fixture canopy or other equipment for control of the fixture or equipment.

quirements, manufacturers are now advertising and describing their switches in terms of ratings for conditions of use—such as rated amps and volts for use with tungsten lamp loads and for use with fluorescent lighting (an inductive load).

Interchangeable Units

As in the case of plug receptacles, wiring switches are made in a line of smaller, light-duty units called interchangeable devices. As many as three of these devices may be installed in the same space required for a single standard-line wiring switch. The largest switch in this line is rated 10-amp for 125 or 250 volts. But these units should not be used where frequent operation of the switch mechanism is required. Units are made in single and double pole types, 3-way and 4-way types, with regular toggle handle or with locking-type mechanism. Three switches or a combination of three units of switches, receptacles and pilot lights may be assembled on a single supporting strap yoke and mounted in a single gang outlet or switch box.

Combination Line

A combination line of wiring switches is very much the same as the

interchangeable line, except that the combination of three units—switches, receptacles and pilot lights—is made in one molded composition body, rather than individual interchangeable units mounted on a yoke. In the combination line, any given combination is fixed by construction. Combination units can be mounted in a single box.

APPLICATIONS of wiring switches vary widely depending upon the type of switches used and the switching functions to be performed. Typical switch applications are as follows:

• **Single-pole switches** are used for controlling lights or appliances from a single point. Such switches are connected in the "hot" leg of the circuit. In cases where the controlled outlet is out of sight from the switch location, a pilot lamp is commonly used and connected across the circuit conductors on the load side of the switch, to indicate if the light is "ON" or "OFF." Such cases include basement or garage lights controlled from a switch in the kitchen of a home, refrigerated-room lights controlled from outside the room and many industrial applications where lights are located in closed rooms and controlled from outside.

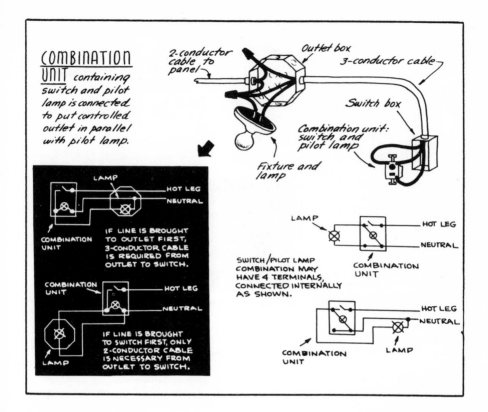

COMBINATION UNIT containing switch and pilot lamp is connected to put controlled outlet in parallel with pilot lamp.

2-conductor cable to panel

Outlet box

3-conductor cable

Switch box

Combination unit: switch and pilot lamp

Fixture and lamp

LAMP

HOT LEG

NEUTRAL

COMBINATION UNIT

IF LINE IS BROUGHT TO OUTLET FIRST, 3-CONDUCTOR CABLE IS REQUIRED FROM OUTLET TO SWITCH.

COMBINATION UNIT

HOT LEG

NEUTRAL

IF LINE IS BROUGHT TO SWITCH FIRST, ONLY 2-CONDUCTOR CABLE IS NECESSARY FROM OUTLET TO SWITCH.

LAMP

LAMP

HOT LEG

NEUTRAL

SWITCH/PILOT LAMP COMBINATION MAY HAVE 4 TERMINALS, CONNECTED INTERNALLY AS SHOWN.

COMBINATION UNIT

HOT LEG

NEUTRAL

COMBINATION UNIT

LAMP

• **Double-pole switches** are also used to control lights from a single point, but they are connected to open or close both of the circuit conductors to the controlled outlet.

• **Three-way switches** are widely used in all types of wiring systems to provide control of lights from two different locations. Typical of this application is the hall light in a home, controlled from either of two switches—one on the first floor and one on the second floor. A light controlled by three-way switches may be turned "ON" or "OFF" from either switch. Again, when the light is out of sight from a switch location, a pilot light may be used with three-way switches to indicate whether the light is "ON" or "OFF."

Proper connection of three-way switches is important to safe operation.

Three-way switches are single-pole switches and must be connected so that switching is down only in the hot circuit conductor. The grounded circuit conductor is carried directly to the outlet to be controlled. It is possible to connect three-way switches with both the hot and grounded circuit conductors connected to each switch. In such a hookup, the controlled outlet is isolated from the supply to provide the "OFF" control. Such a hookup is wrong and dangerous. Failure of the switch mechanism could cause a short circuit.

• **Four-way switches** are used with two three-way switches to provide control of lights from three or more different points. The four-way switches are connected in the hot leg of the circuit to the controlled outlet, with a three-way switch connected on each side of the one or more four-way switches. A

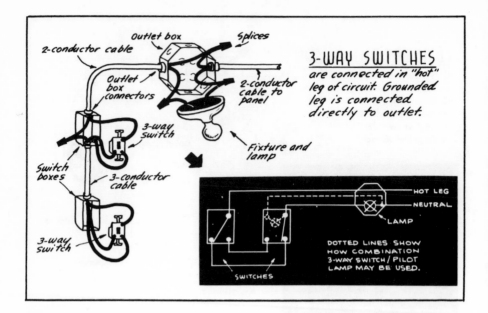

3-WAY SWITCHES are connected in "hot" leg of circuit. Grounded leg is connected directly to outlet.

Outlet box

2-conductor cable

Splices

Outlet box connectors

2-conductor cable to panel

3-way switch

Switch boxes

3-conductor cable

3-way switch

Fixture and lamp

HOT LEG

NEUTRAL

LAMP

DOTTED LINES SHOW HOW COMBINATION 3-WAY SWITCH / PILOT LAMP MAY BE USED.

SWITCHES

single four-way switch connected between two three-way switches will provide three points of control. For each additional point of control required, another four-way switch must be connected in series between the three-way switches. To control a light from, say, five different locations, requires two three-way switches with three four-way switches connected between them.

• **Three-pole and four-pole** switches are made in tumbler and rotary types for switching applications involving three or four circuit wires.

Wiring Boxes

The term "wiring devices" includes the wide variety of small boxes used in wiring systems. These are the boxes used at outlets, switch locations and junction points. As used here, the word "outlet" describes any point in a wiring system at which electric power may be taken to supply lighting fixtures, motors, fixed appliances, cord connected devices and appliances and any other current-consuming equipment. Although generally constructed

of steel—for armored cable and steel conduit systems, outlet boxes are also made of non-metallic material for use with non-metallic-sheathed cable systems, open wiring on insulators and concealed knob-and-tube work.

In general, wiring boxes are divided into several categories according to construction and application, as follows:

Outlet Boxes—These are the boxes used at outlet points in wiring systems. They are made in several cross-section shapes—octagonal, round, square and rectangular.

Octagonal and round outlet boxes are generally used for ceiling outlets in all types of concealed wiring systems in all types of construction. Round boxes are not used where conduit must enter the box from the side because of the difficulty in making a good connection on a rounded surface. Octagonal boxes are also used for wall outlets to lighting fixtures.

Both round and octagonal boxes are made with various combinations of knockouts in sides and bottom to accommodate use with almost any configuration of conduit runs. These boxes are available with built-in clamps for holding armored cable or non-metallic-sheathed cable inserted in the box. Built-in fixture studs or provisions for mounting separate fixture studs are also standard parts of such boxes. Many types of covers are made for these boxes.

Special octagonal boxes are made for concrete work. These boxes are made in a range of depths to 6 inches, with removable back plates and knockouts for conduit connections in both the sides and back plate.

Square and rectangular or oblong outlet boxes find application for wall outlets in conduit wiring systems when the boxes are embedded in masonry or set in brick or tile. Both boxes are provided with combinations of knockouts and are also used in other types of building construction. Square boxes have built-in fixture studs or provisions for mounting separate fixture studs.

Oblong boxes are commonly called "gang" boxes, a term which refers to the fact that a number of standard wiring devices can be mounted in the box. A four-gang box, for example, will accommodate four standard wiring devices mounted side-by-side in the box.

A wide variety of covers are made for square and oblong boxes, with openings and lugs to accommodate flush wiring devices. These box covers have raised lips around the main opening, which, when the box is mounted slightly back from the surface of the wall, bring the devices flush with the wall surface. Installed, the box and its cover are plastered over up to the lip of the main opening so that the flush plate which is finally mounted over the device will cover the remaining opening. A different group of covers is available for square boxes accommodating flush wiring devices in exposed wiring systems.

I N ADDITION to outlet boxes, electric wiring systems make use of other types of boxes, as follows:

Sectional Switch Boxes—These boxes are used in concealed wiring systems in which the boxes are not embedded in masonry.

Sectional boxes have removable sides to allow ganging of a number of boxes side-by-side to accommodate several wiring devices at a single location, in a compact housing. These boxes do not require covers; they accommodate standard flush wiring devices with standard flush plates.

The boxes have a number of conduit knockouts and may have internal clamps for holding armored cable or non-metallic sheathed cable.

Utility Boxes—These boxes are designed for use in exposed conduit wiring systems. They have knockouts in bottom and sides to accommodate a variety of conduit runs.

Flush Plates

Flush plates are the finishing covers used on flush wiring devices in concealed wiring installations. Flush plates have the necessary small cutouts to accommodate toggle handles, pilot lights, convenience receptacles. They give the installation a neat appearance.

Flush plates are made in one-, two- and three-gang sizes—in various combinations of openings—to meet the requirements of almost any combination of devices.

They are made of bakelite, plastic, steel, enameled metal, brass and other materials for special applications.

Box Supports

To mount and support various types of outlet boxes and wiring boxes in the many types of construction used today, a number of box support devices are available.

Mounting straps of flat steel stock

are made in a number of types and sizes with and without attached fixture studs. Made in different lengths, depending upon the type of box to be supported, typical straps or bar hangers are 18-, 21-, and 24-in long, with holes at both ends of the bars for nailing to joists or studs.

Units without fixture studs have slots in their middle sections to accommodate stove bolts holding boxes. These are known as box cleat hangers. Some straps are straight; others, known as offset hangers, have offset sections in their middle to accommodate mounting boxes of different depths so that boxes may be recessed in ceilings or walls.

A variety of types of mounting straps are made to meet all job conditions. Boxes and strap hangers are also available in preassembled combinations of correct hangers for the different boxes. Assemblies are made for armored cable, non-metallic sheathed cable, thinwall and rigid conduit systems.

Other types of box support devices include footed box support bars for wall mounting boxes by attaching the box by screws through its side to a slot in the vertical support bar.

There are also special supports for mounting switch boxes to beaverboard or sheetrock or holding boxes installed in walls of old houses where no other support can be obtained for the box, and flat bar hangers for mounting octagonal boxes in hung ceilings.

Insulated Devices

A wide range of standard wiring boxes, covers and plates are made of insulating materials—like porcelain, bakelite, etc.—to provide insulated wiring systems of standard concealed knob and tube wiring, open cleat wiring or non-metallic sheathed cable.

Grounding of device enclosures is not required because the box material is an insulator. Providing ease of installation and economy, these units assure safety to personnel if a hot conductor should ground to the device.

Use of metallic boxes with non-metallic wiring systems and without a grounding conductor carried to the boxes and connected to them presents the hazard of shock if a hot conductor should become common to the box.

Other Boxes

In addition to the boxes described, there are other boxes for special applications. Typical of these is the floor box. Floor boxes are made to accommodate plug receptacles, to provide for signal circuit outlets and to serve as junction boxes for conduit wiring systems embedded in concrete floors or concealed in wood floors. Units are made in adjustable and non-adjustable types.

The adjustable boxes have tops which can be adjusted by means of leveling screws over a range of about $\frac{1}{2}$-in for vertical positioning of the unit to meet the conditions of the floor. These units can also be adjusted for about 15° of angular positioning. Units are made in single and gang outlet types.

Another type of wiring box is the door switch box, made to provide mounting of switches in door jams.

Pull boxes and junction boxes are used in conduit raceway systems for a number of reasons. Pull boxes, for instance, are used in long conduit runs which contain more than the allowable number of bends. The code limits the number of bends allowed in a conduit run, and the practical allowable number will depend upon the physical conditions which will allow conductors to be pulled into the conduit.

The length of the run, the size of the conduit and the number and types of conductors to be pulled will determine limits for bends.

Generally, two or three equivalent 90° bends represents a maximum between pulling points. Pull boxes permit

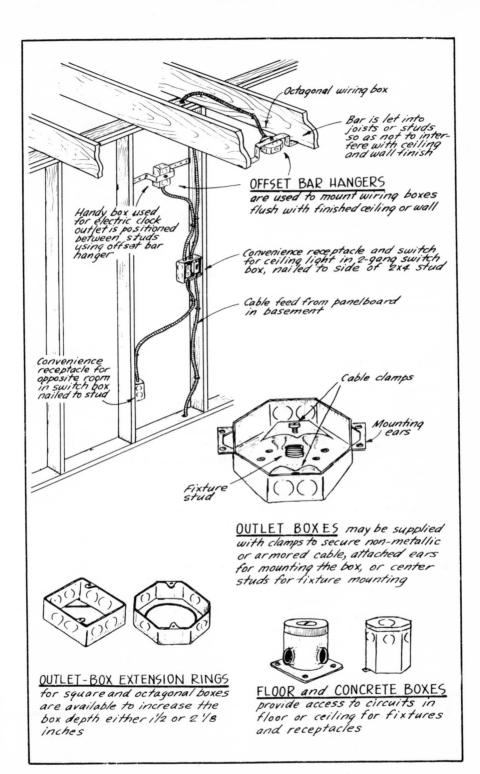

Octagonal wiring box

Bar is let into joists or studs so as not to interfere with ceiling and wall finish

OFFSET BAR HANGERS are used to mount wiring boxes flush with finished ceiling or wall

Handy box used for electric clock outlet is positioned between studs using offset bar hanger

Convenience receptacle and switch for ceiling light in 2-gang switch box, nailed to side of 2x4 stud

Cable feed from panelboard in basement

Convenience receptacle for opposite room in switch box nailed to stud

Cable clamps

Mounting ears

Fixture stud

OUTLET BOXES may be supplied with clamps to secure non-metallic or armored cable, attached ears for mounting the box, or center studs for fixture mounting

OUTLET-BOX EXTENSION RINGS for square and octagonal boxes are available to increase the box depth either 1½ or 2⅛ inches

FLOOR and CONCRETE BOXES provide access to circuits in floor or ceiling for fixtures and receptacles

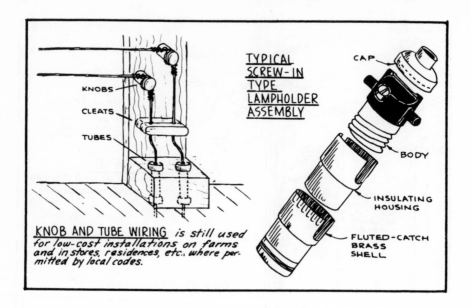

TYPICAL SCREW-IN TYPE LAMPHOLDER ASSEMBLY

KNOBS

CLEATS

TUBES

CAP

BODY

INSULATING HOUSING

FLUTED-CATCH BRASS SHELL

KNOB AND TUBE WIRING *is still used for low-cost installations on farms and in stores, residences, etc., where permitted by local codes.*

pulling of conductors to the box and from the box to the next point of pull, and they permit ready change in direction of the conduit run. Such boxes should be long enough to provide adequate radius of bending for the conductors.

Junction boxes are similar to pull boxes and are used for the same general purposes—eliminating bends in a large number of parallel conduit runs, providing change of run direction, providing taps into conductors, providing for conduit transpositions, etc.

Pull and junction boxes are made in many sizes, with dimensions up to multiple feet and with wide varieties of sizes and locations of knockouts for connecting conduit. Such boxes are generally made of standard gauge sheet steel, with screw covers, but many cast iron boxes are made for heavy-duty applications.

Another type of common wiring device is the lampholder or lamp receptacle.

This device is familiar to everyone as the type of lamp socket assembly

used around the home—in garages, in attics, in basements—and in many commercial and industrial areas where lighting requirements are economically satisfied by wall or ceiling mounted bare incandescent lamps in surface type metal or porcelain socket assembly. These devices are made in several types for different applications.

In general, a typical lampholder is an assembly which accepts a screw-in type incandescent lamp. Such a device consists of a screw-threaded lamp socket, mounted in a protective and insulating housing, with some provision for supporting the whole assembly in its location.

Types of Lampholders

One of the oldest and most widely used types of lampholder is the fluted catch, interchangeable **brass shell type.** Such a device is made up of two basic parts: 1) a body, which is a short cylindrical housing with a lamp socket and with or without a switch mechanism for operating the lamp; 2) an attachment part to which the lampholder body can be readily attached, which may be either a cap to allow mounting of the lampholder on conduit, or a pendant cord or a base to permit surface mounting of the lampholder on an outlet box.

Typical bodies may incorporate a push-type switch, a pull chain switch or a rotating key-type switch. They may have no switch at all. Any type of body may be used with any type of cap or base, provided, of course, the two connecting parts have attachment provisions of the fluted-catch type.

A more recently developed type of interchangeable . lampholder is the **threaded-catch type.** Basically, this device is similar to the older fluted-catch lampholder.

But attachment of the body to a cap or base is made by means of a threaded ring, a third piece, which holds the cap and screws onto the threaded portion at the top of the shell. The bodies and caps and bases of threaded-catch lampholders may be made of composition or they may have brass outer shells.

Still another type of catch arrangement which is used with an interchangeable line of lampholders is the **snap catch.**

In the porcelain snap-catch line, the bodies are attached to caps or bases by means of two bayonet hooks in the body which engage phosphor-bronze flexible contact catches in the cap or base. The parts are securely held in this manner, but may be separated easily when necessary.

In these devices, the hooks and contact catches provide electrical connection between the parts, in addition to mechanical connection. This porcelain line of lampholders includes the same general types of caps, bases and bodies as the types in the fluted-catch line. These porcelain devices are used in places with corrosive atmospheres.

For the same applications, another

line of porcelain lampholders is available, using two screws to hold a body and cap or base together. Hard composition interchangeable lampholders are also available.

Non-interchangeable types of lampholders are made in many types, sizes and constructions to meet a wide range of lighting applications. These devices are generally made of porcelain, plastic or composition.

Typical units are made for mounting on outlet boxes, for mounting in lighting fixture canopies and for surface mounting in exposed wiring installations.

Lampholders are made for candelabra, intermediate medium screw and mogul base lamps. A widely used type of non-interchangeable lampholder is the weatherproof type. This type is used for outdoor and wet-location applications and may be made of one-piece porcelain, composition or rubber, with the leads sealed into the housing of the assembly.

Insulators

For interior wiring installations where conductors are not installed in raceways, there are a number of types of insulators used. Although open wiring is not as widely used today as it was years ago, it still finds advantageous application in some commercial and industrial areas.

The most common types of insulators used for open wiring are: knobs, cleats, tubes and crane and rack insulators. These insulators are made of porcelain and may use nails or screws for mounting to wood joists, beams, studs, etc.

For outside electrical distribution, overhead and underground, there are many types of insulators. These include:

• Suspension insulators for supporting conductors on aerial transmission lines.

• Pin-type insulators for supporting crossarms and dead-ending aerial conductors.

• Spool and wire-holder type insulators for use singly or on rack assemblies for supporting low-voltage circuits on the sides of poles or on buildings for dead-ending service wires.

• Strain insulators for use in guy wire assemblies and in line wires at dead-end points.

Wiring devices for outside distribution cover a very wide range of other parts and hardware.

Of course, there are many other types of wiring devices than those covered. There are many types of drop cords, extension cords, lamp shade holders, rosettes, socket adapters, plugs, etc. But the more important types of devices have been covered.

Signals

IN ALL TYPES of modern buildings—homes, schools, office buildings, industrial plants—electrical signal devices and systems find wide and important application. Such signals include both audible and visual types, either manual or automatic.

Door Signals

Door signals are by far the most common type of signal in general use. These include bells, buzzers and chimes. In home applications, door signals are operated by means of pushbuttons placed outside of the entrance doors. These signals are connected to low voltage circuits— typical ac voltages are 8-12, 24 and 48 volts; typical dc voltages are 8 and 24 volts—and are operated by their associated momentary-contact, pushbutton switches which are connected in series with the devices.

In bells, buzzers and chimes, the

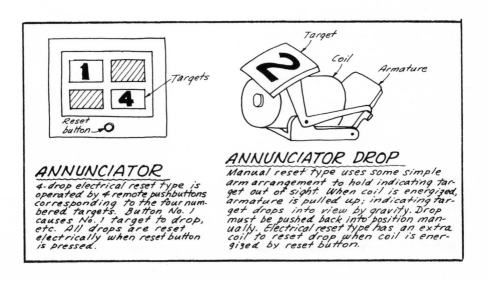

ANNUNCIATOR

4-drop electrical reset type is operated by 4 remote pushbuttons corresponding to the four numbered targets. Button No. 1 causes No. 1 target to drop, etc. All drops are reset electrically when reset button is pressed.

ANNUNCIATOR DROP

Manual reset type uses some simple arm arrangement to hold indicating target out of sight. When coil is energized, armature is pulled up; indicating target drops into view by gravity. Drop must be pushed back into position manually. Electrical reset type has an extra coil to reset drop when coil is energized by reset button.

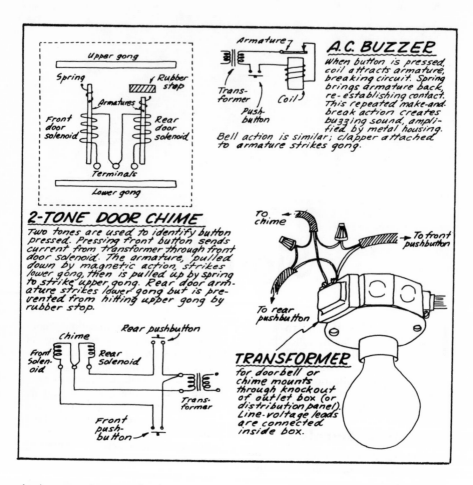

A.C. BUZZER

When button is pressed, coil attracts armature, breaking circuit. Spring brings armature back, re-establishing contact. This repeated make-and-break action creates buzzing sound, amplified by metal housing. Bell action is similar; clapper attached to armature strikes gong.

2-TONE DOOR CHIME

Two tones are used to identify button pressed. Pressing front button sends current from transformer through front door solenoid. The armature, pulled down by magnetic action, strikes lower gong, then is pulled up by spring to strike upper gong. Rear door armature strikes lower gong but is prevented from hitting upper gong by rubber stop.

TRANSFORMER

for doorbell or chime mounts through knockout of outlet box (or distribution panel). Line-voltage leads are connected inside box.

basic operating mechanism is an electromagnetic solenoid with an operating arm attached to its armature. In bells and buzzers, operation depends upon making and breaking of the coil circuit to produce an oscillatory motion of the armature and the arm attached to it. To accomplish this end, the solenoid assembly is equipped with a normally closed set of contacts in series with the coil of the solenoid. When the solenoid is energized, the armature is moved against the restraining force of a spring. This movement of the armature opens the normally-closed contacts and breaks the circuit to the solenoid coil. The spring then returns the armature to its at-rest position. But as soon as the armature is back in this position, the contacts are closed again, and the circuit is completed through the solenoid coil, energizing it. The action then goes through the cycle again. And so long as the pushbutton is held depressed, the oscillatory or vibrating motion will continue.

In the case of a bell, the arm attached to the armature is equipped with a small hammer which follows the vibrating motion and strikes the sounding gong. In the case of a buzzer, the vibrating action of the armature parts sets up a buzz which is mechanically amplified by the metallic housing of the unit.

Door chimes are also operated by means of solenoids, but do not include the vibrating assembly. In the simpler types of chime assemblies, a short wooden striker is set in the one or more armatures. When the coil circuit is energized, the armature moves, carrying its striker against a tuned metallic sounding board. Depending upon the particular unit, there may be one or two audible tones produced by the action of the striker. Some door chime assemblies have two separate solenoids—one which produces one tone when energized and another which produces two tones when energized. The tones thus distinguish between two circuits which are individually operated from two different pushbuttons at two different locations. This is the common hookup used in homes to distinguish between door signals from the front and back door.

Circuits for the foregoing equipment are simple two-wire hookups, powered from small signal transformers which are rated 120-volt primary to the low-voltage secondary. The transformer in any system is generally mounted at or close to the service entrance equipment layout. Connections on the secondary side of the transformer—from transformer to door pushbuttons to signal units—are made with bell wire which does not have to conform to the wiring techniques for power wiring.

Complex door chime systems—such as the hookups which produce six chime tones when the cycle is initiated by one push of a momentary-contact pushbutton—usually consist of the striker assembly at the chimes and an operating panel mounted near the service equipment or in some accessible place in the basement or utility area. This operating panel generally contains the bell transformer to supply the low voltage, a motor-operated timing mechanism and a relay to operate the motor.

In homes, bells, buzzers and chimes are used in manual systems. Somebody must push the button to make the signal operate. In many commercial and industrial buildings, however,

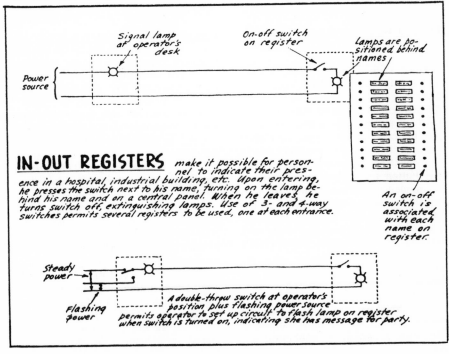

IN-OUT REGISTERS *make it possible for personnel to indicate their presence in a hospital, industrial building, etc. Upon entering, he presses the switch next to his name, turning on the lamp behind his name and on a central panel. When he leaves, he turns switch off, extinguishing lamps. Use of 3- and 4-way switches permits several registers to be used, one at each entrance.*

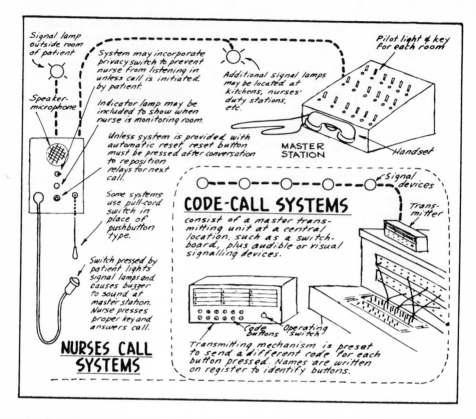

Signal lamp outside room of patient

System may incorporate privacy switch to prevent nurse from listening in unless call is initiated by patient.

Indicator lamp may be included to show when nurse is monitoring room.

Speaker-microphone

Additional signal lamps may be located at kitchens, nurses' duty stations, etc.

Pilot light & key for each room

MASTER STATION

Handset

Unless system is provided with automatic reset, reset button must be pressed after conversation to reposition relays for next call.

Some systems use pull-cord switch in place of pushbutton type.

Switch pressed by patient lights signal lamps and causes buzzer to sound at master station. Nurse presses proper key and answers call.

CODE-CALL SYSTEMS

consist of a master trans-mitting unit at a central location, such as a switch-board, plus audible or visual signalling devices.

Signal devices

Trans-mitter

Code buttons   Operating switch

Transmitting mechanism is preset to send a different code for each button pressed. Names are written on register to identify buttons.

NURSES CALL SYSTEMS

the door signal is put on an automatic basis to respond to movement of the door.

Door signals in apartment houses and in other applications where protection against entry is required frequently include electric door openers. In these systems, the caller presses a pushbutton located outside a locked main entrance door. In the case of apartment house systems, there is one pushbutton for each tenant. A signal is sounded in the apartment. The tenant then presses a button in the apartment to release the lock on the main entrance door. The signal bells and buzzers and the electrically-operated door locks in these systems are operated from a low-voltage dc or ac supply.

Annunciators are signal devices which provide visual indication of a call from a particular pushbutton station and usually include a buzzer or bell for calling attention to the visual indicator. A typical system is made up of the widespread pushbutton stations connected into the annunciator panel. This panel contains a number of small windows which normally are blank. When one of the system pushbuttons is pushed, a solenoid moves an indicator into the window associated with the pushbutton. This indicator may be a number or lettered description such as "Dining Room", "Vice President" or "Shipping", to indicate the point from which the call originates. Some annunciator panels contain pushbuttons for acknowledging calls by returning a signal to operate a bell or buzzer at the originating station.

Annunciators are made for operation on low voltage and some systems are made for use at 120 or 250 volts. The drop indicators in annunciators may be reset to their normal positions by manual mechanisms or electrically.

Annunciators are used in large residences to provide communication between residents and the help, in hotels and restaurants for the same reason and in commercial and industrial buildings where one person has to respond to calls from a number of different locations.

PAGING SYSTEMS of the visual and audible type, other than voice call sound systems, find wide application in modern industrial plants, commercial buildings, schools and hospitals. Systems of this type are used to call and locate individuals for answering telephones, attending meetings, providing service, etc.

Coded type call systems consist of a centrally located electric code transmitter unit which is controlled to send coded energy pulses to signal sounding devices. These devices may be bells, horns, chimes or low-level tone sounders. In such systems, personnel in the building are assigned individual code signals. These systems are used where calls will originate from a central control station.

Call systems of almost any type and complexity can be made up from circuits using pushbuttons and buzzers. These are common in office occupancies. The pushbutton is placed at a desk or other work area, and the buzzer is placed at the location of a person (secretary, for instance) who will respond to the signal buzz by prearrangement. Or a pushbutton and buzzer can be placed at each of the two points between which call will be made, thereby providing return signal or call-back ability. In such hookups, the pushbutton acts as a switch to close the circuit which supplies low-voltage energy to the buzzer. Small bell transformers (generally under 50 watts) are used to supply the low voltage. Electrically operated door locks, the same as the ones used for apartment houses, are commonly combined with buzzer call systems in offices and security locations such as laboratories.

A special type of call system is the lecture call system used in school auditoriums and other large assembly halls. The purpose of such a system is to provide a return call method between the stage, where a lecturer is talking, and the projection booth in the rear of the auditorium, where an operator is coordinating slide projections or movies with the lecturer's presentation. The system is made up of a buzzer and pushbutton at each location, and sometimes includes a pilot light in the booth. The pushbutton at the stage end is commonly a pendant type on a long portable cord plugged into an appropriate outlet at the side or edge of the stage.

Visual call systems used in many commercial and industrial areas consist of keyboards or pushbutton stations which are used to operate annunciator drops, flashing . lamps or selective buzzer signals. These systems may also combine return call to pilot lights on the originating panel. Depending upon the particular installation, a code of numbers is set up for personnel involved.

Hospital Calls

A variety of call systems is commonly used in hospitals, including: visual and audible nurse-call systems, emergency call, psychopathic alarm call, intern call, ambulance call, nurses' home call and door calls.

Nurse-call systems are used by hospital patients to call a nurse to the bedside. By pressing a pushbutton at his bed, the patient causes a light to come on in the corridor over the door to his room, in duty rooms and diet kitchens and at the nurses' station for the floor or section. In wards, a pilot light at the calling bed comes on when the button is pressed. At the nurses' station, there may be a single common lamp which signals in response to any pushbutton or

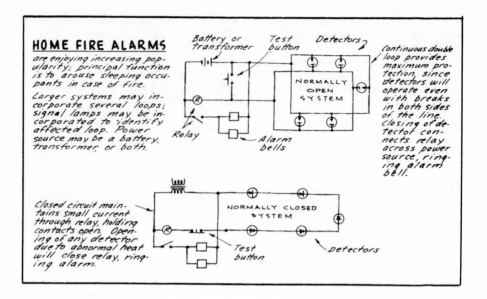

HOME FIRE ALARMS

are enjoying increasing popularity; principal function is to arouse sleeping occupants in case of fire.

Larger systems may incorporate several loops; signal lamps may be incorporated to identify affected loop. Power source may be a battery, transformer, or both.

Battery or transformer
Test button
Detectors

NORMALLY OPEN SYSTEM

Relay

Alarm bells

Continuous double loop provides maximum protection, since detectors will operate even with breaks in both sides of the line. Closing of detector connects relay across power source, ringing alarm bell.

Closed circuit maintains small current through relay, holding contacts open. Opening of any detector due to abnormal heat will close relay, ringing alarm.

NORMALLY CLOSED SYSTEM

Test button

Detectors

there may be an annunciator with a lamp for each room. Some systems also include a supervisory annunciator at the chief nurse's quarters to cover all floors.

Two-way talk facilities are commonly combined with visual nurse-call systems to enable the nurse at the station to converse with the patient. This system includes not only lamps and buzzers on low voltage circuits but also microphones and loudspeakers with controls for selective talking to individual rooms.

Emergency call is a feature which may be added to any nurse-call system. With this system, a nurse who has responded to a call is able to summon assistance by means of a separate button which activates a special circuit to light a red emergency light at the station and in the corridor and to sound a continuous bell. A variation on this is the psychopathic alarm system which enables an attendant to summon help to a particular room when necessary.

Paging systems of the visual and audible type are used to locate doctors and other staff personnel throughout the hospital. (Of course, audio voice systems accomplish this very effectively also.) The visual systems are annunciator types with attention buzzers at the indicating stations. The audible systems use bells, buzzers or other sounders operated on a coded basis to designate specific persons. Typical systems are controlled from a central point in the hospital or may be designed for signalling from a number of stations. In these systems, the signal devices are located throughout the building at points where they can most effectively perform their jobs.

Intern call systems are generally of the return call type (pushbutton and buzzer at the central control station in the hospital and in the room of each intern), or are master telephone systems. Some return call systems include indicating drops at the rooms and at the central point.

Ambulance call systems provide a means for indicating to the central control station that a patient is being brought in at the ambulance entrance. These are generally annunciator type systems operated by a pushbutton at the ambulance entrance.

Register systems, common in hospitals for indicating the presence of a doctor in the building, are located at building entrances and tied in to a central indicating panel, usually at the telephone operator's position. Upon entering, the doctor presses a switch opposite his name on the panel, illuminating his name on that panel and on the central panel. When he leaves, he turns off the switch, extinguishing the signal lamps.

A "recall" feature provides means for the operator to throw a switch opposite a particular name when she has a message for the individual. When he enters and presses his switch, the lamp behind his name flashes on and off, and he calls the operator for the message. Registers may be used in multiple, using 3- and 4-way switches, permitting a panel at each entrance.

SIGNAL equipment layouts in all types of modern buildings commonly include one or more alarm systems. Typical of such systems are: fire alarms, burglar alarms, door alarms, hold-up alarms, sprinkler alarms and transformer alarms and other industrial-type alarms.

Fire Alarms

Fire alarms are used in industrial, institutional and commercial buildings to warn occupants of the presence of fire. The use of fire alarms in individual residences is gaining widespread acceptance.

Fire alarm systems may be of the non-coded or coded type and operation may be manual, automatic or a combination of both. A typical manual fire alarm system is made up of a central control panel supplied by a circuit which energizes the system, operating stations and electrically operated gongs to sound alarm. The operating stations (such as "Break glass and pull lever") provide points at which the alarm may be initiated.

Non-coded fire alarm systems transmit only a general alarm and do not indicate the location of the operated stations. In coded systems, the signal distinguishes the station

which was operated. Non-coded systems are best applied in small areas; coded systems, in large areas or multiple buildings.

In system layouts, operating stations are placed in paths of escape and spaced so there is one within 100 feet of any point. The sounding devices—generally bells, occasionally projector horns—are spaced to provide ready hearing from any point.

Automatic fire alarm systems use detectors to sense dangerous temperatures and to actuate alarm or to accomplish other actions to minimize the hazard—such as disconnecting blowers which would create draft or closing dampers in ducts to prevent drafts. The thermostatic detectors in such systems act as switches in the circuit, closing or opening on dangerous temperatures. Automatic systems are especially important in areas which are left unattended, such as storerooms and warehouses. They are commonly combined with manual systems.

Fire-line alarm systems are similar to standard coded systems with the addition of manually operated signal stations installed on alternate floors of tall buildings. Each of these stations has a hinged, locked cover and is designed to be used only by fire department personnel to transmit signals to the pump room. Fire-line telephone systems which are also commonly used in tall buildings provide for voice intercommunication between the pump rooms and individual floors of the building.

Burglar Alarms

Burglar alarm systems for use in modern buildings consist of signal sounding devices—bells, buzzers, etc. —connected on circuits which are energized by devices which detect unauthorized entry into a building or area. Such systems are used to protect doors, windows, elevator openings, skylights, etc.

The sensing devices include: various types of spring contacts, switches,

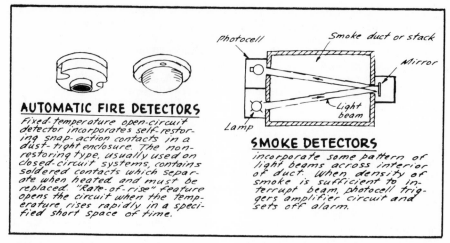

AUTOMATIC FIRE DETECTORS

Fixed-temperature open-circuit detector incorporates self-restoring snap-action contacts in a dust-tight enclosure. The non-restoring type, usually used on closed-circuit systems, contains soldered contacts which separate when heated and must be replaced. "Rate-of-rise" feature opens the circuit when the temperature rises rapidly in a specified short space of time.

SMOKE DETECTORS

incorporate some pattern of light beams across interior of duct. When density of smoke is sufficient to interrupt beam, photocell triggers amplifier circuit and sets off alarm.

foil tape on glass doors or window and network of wired lattices. Photoelectric devices are also used as detecting devices to initiate an alarm upon intrusion. Depending upon the type of building or area being protected, the signal sounding devices may be placed at the location to provide general alarm or in a guard's office to provide discreet alarm. Such systems may also be tied into police headquarters.

Door Alarms

These are similar to burglar alarms. A typical system consists of contact devices at the doors to be supervised, circuits to signal lamps and/or a bell or buzzer at the guard's station or other supervisory location and, frequently, a large alarm bell adjacent to each guarded door.

Hold-Up Alarms

Alarms of this type are commonly used at disbursing offices, cashiers' locations, tellers' windows in banks and other locations where large amounts of money are handled and where holdups might occur. The circuit usually consists of a standard or special pushbutton—such as foot or knee operated—at the counter or desk to provide discreet initiation of an alarm signal.

The signal indicating devices may be audible, visual or both and may be placed at any number of locations to notify guards and/or police and to facilitate apprehension of the thieves. The alarms in these systems are generally operated discreetly.

Sprinkler Alarms

These alarms are arranged to signal building operating personnel when sprinkler heads open, when serious leaks occur or when other parts of the sprinkler system malfunction. Valves are equipped with contacting devices to initiate transmission of a signal to an audible or visual indicating device at a supervisory point.

Transformer Alarms

Alarm systems may be hooked up to supervise operation of transformers, automatically initiating an audible or visual alarm when temperature and pressure of oil-cooled transformers reaches a dangerous level. Switching devices are located in the transformer housing. Provision may also be made for the release of CO_2 in the event of fire.

Smoke Detection

Detection systems for smoke control consist of photoelectric sensing

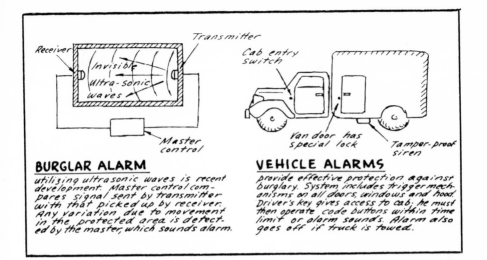

BURGLAR ALARM

utilizing ultrasonic waves is recent development. Master control compares signal sent by transmitter with that picked up by receiver. Any variation due to movement in the protected area is detected by the master, which sounds alarm.

VEHICLE ALARMS

provide effective protection against burglary. System includes trigger mechanisms on all doors, windows and hood. Driver's key gives access to cab; he must then operate code buttons within time limit or alarm sounds. Alarm also goes off if truck is towed.

equipment installed in the smoke stack, with the signal output on dense smoke used to warn operating personnel or to automatically operate other equipment to correct the bad smoke condition. Similar systems can be used in ventilating ducts and used as fire alarm equipment, with the output of the photoelectric circuit operating signal sounding devices. In both cases, the signal is initiated when the smoke becomes dense enough to break or sufficiently weaken the light beam which falls on the photo cell to establish normal condition for the circuit.

Many other types of specific signal systems are available, consisting of one or more sensing devices which respond to some abnormal condition by actuating one or more warning devices.

Clock Systems

ONE OF THE FASTEST growing applications of indicating and signalling equipment in modern industrial, commercial and institutional buildings is the clock and program system. Basically, such systems consist of a number of electric clocks spread throughout the building and controlled, for accuracy of time, by special control circuits or by electronic wireless means from a central control assembly which also operates signal bells or horns according to a preset schedule.

Application

Typical application of clock and program systems is as follows:

Industrial Plants — Systems are used for indicating the time of day and operating bells or horns at pre-determined times, such as at starting and stopping of work, for rest periods, lunch periods, etc.

Commercial Buildings — Clock and program systems for time indication and programmed signalling are commonly used throughout all or part of a building.

Schools — All types of schools, colleges and other institutions of learning or instruction make extensive use of clock and program systems. Clocks

ELECTRONIC CLOCK SYSTEM

National Bureau of Standards radio signals, transmitted every hour on the hour, are received by system antenna and synchronize master clock. This clock has supervising contacts which control high-frequency generator output. The controlled generator tone, superimposed on ac line, keeps local clocks on time. All clocks operate on 115-volt branch circuits. Capacitors isolate 60-cycle voltage from clock controls.

Antenna
Lightning arrester
Generator
Local clocks
Branch circuit panel
Capacitors
AC power line
Master clock
Generator control unit

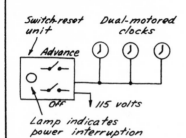

SWITCH-RESET CLOCK SYSTEM

Switch-reset unit

Advance

Dual-motored clocks

Off

115 volts

Lamp indicates power interruption

All clocks are driven normally by synchronous motors. If power failure causes clocks to lose time, "advance" switch is thrown, energizing second motor in clock, which advances clock at rapid rate until lost time is made up. "Off" switch permits stopping clocks to change from daylight savings to standard time. Automatic reset systems use control unit which records length of outage, then automatically advances clocks to correct time.

are mounted in the classrooms and/or corridors to indicate correct time, and signal devices are strategically located throughout the premises to sound the beginning or end of various periods of the day.

Hospitals — Clock systems are very important in hospitals where they are used for general timekeeping and for specific time scheduling as in the administration of anesthetics. Clocks are located in corridors, offices, nurses' stations, operating rooms, laboratories, etc. For special timing needs, seconds-beat clocks should be installed in operating and delivery rooms.

Hotels — Clocks should be installed in lobbies, offices, engine rooms, kitchens, ballrooms, dining rooms and grilles.

There are several different operating arrangements used in modern clock systems. The common types of systems are as follows:

Synchronous wired dual-motored clocks, central control system — In this system, each clock contains two motors. In normal operation, one motor operates the clock to keep accurate time. The second motor in each clock operates to correct the time indication of the clock when some abnormal condition has caused the clock to stop, as in the case of a power failure. When power is restored to an interrupted circuit, all clocks start again automatically but

are slow by the time duration of the outage. The second motor in each clock, called the "reset" motor, is then operated from the central control panel to advance the clocks at a greatly accelerated rate to correct time indication.

Reset motors in the above clock systems may be actuated from the central control station in two ways: by manual key switch or by automatic reset control.

In the first case, the reset control is equipped with a pilot lamp to indicate power outage and with two key operated switches, one for advancing the clocks to reset the correct time and the other for stopping the clocks, as might be required for time changing or for repairs.

In the second case, the central station contains control equipment which will automatically advance all clocks to correct time upon restoration of power after an outage (up to 12 hour duration). This control assembly has reserve power to operate a number of the clocks during periods of power outage, and is equipped with switches for making time changes.

Master time control with synchronous wired clocks — Each clock in this system contains a synchronous drive motor for normal operation of the clock and a special mechanism for time correction. Each clock is individually checked once each hour for accuracy of time, and another

PROGRAM CLOCK SYSTEM

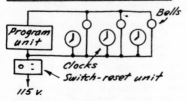

Addition of audible signals and program unit to switch-reset system causes bells to be sounded in accordance with preset program schedule. By using a separate circuit for each bell, they can be made to sound independently of each other at different times.

IMPULSE CLOCK SYSTEM

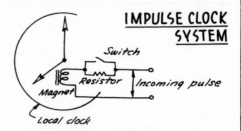

Master clock sends impulses every minute through normally closed switch; drive magnet advances hand one minute with each pulse. Cam opens switch during 59th minute while master clock sends out rapid pulses to advance any slow clocks to the 59th minute. A higher-voltage pulse then advances all clocks to the hour.

---

correction is made once each twelve hours. The checks are made against the master time control at the central control assembly for the system. Variations from correct time are eliminated at each check.

In this system, power outage will stop the clocks, but they will resume operation as soon as power is restored. Correction of time indication will be made within the hour for outages of less than an hour's duration and at a pre-selected hour for power interruptions up to twelve hours.

Unlike the usual electric clock, these minute-impulse clocks have no motor and do not operate on line voltage. A drive magnet is used to advance the hands.

Master time control with minute impulse wired clocks — In this system, each of the clocks is operated by impulses from the master time control panel. These impulses occur each minute to advance the clocks' hands one minute. Each hour, the master control transmits another pulse which provides automatic correction of all clocks. Pulses are also provided to restore correct time indication after power failure.

Electronic controlled synchronous clocks — Each clock in this system is simply connected to a 120-volt ac outlet on a regular lighting and appliance circuit to power its synchronous motor. High frequency pulses are superimposed on the regular power and light wiring system at the central control panel. The pulses are coordinated with a master time control and are sensed by special receiver devices at each clock, providing time correction.

Program control panels are made in a variety of types for simple or complex scheduled operation of audible signalling devices, like bells or horns, or for control of remotely located equipment, like contactors supplying lighting, motors for air conditioning or other power loads. In wired systems, the program panel is installed as part of the clock system and provides pre-scheduled operation of individual signal circuits to meet all requirements for selective control. Program controls for wireless (carrier frequency) systems schedule the transmission of coded pulses for selective signalling and control and provide for supervision of electronic controlled clock systems.

Sound Systems

A NY SOUND SYSTEM is a hook-up of several pieces of equipment which provides a complete facility for paging, public address, announcing and/or music distribution. Typical sound systems encompass a wide variety of equipment and devices. Familar applications include: motion picture sound systems, including rapidly developing stereophonic sound systems; paging systems in airports and railroad stations; public address systems in churches and auditoriums; the office intercom system; and systems for distributing background music in restaurants and stores.

The basic components used in sound systems are separated into three categories: signal sources, amplifiers and loudspeakers. Any sound system—regardless of size, complexity and purpose—must include at least one component from each category. A number of different or similar types of signal sources may be used with one or more amplifiers to feed any number of loudspeakers of various sizes and types.

Signal Sources

In a sound system, the signal introduced into the amplifying equipment originates in one of the following input (signal source) devices:

- **Microphone** — which transforms sound waves into electrical signals and feeds them to the amplifier.
- **Radio tuner**—which picks up AM (amplitude modulated) or FM (frequency modulated) broadcast radio programs and feeds them to the amplifier.
- **Record or tape player**—which picks up the electrical equivalent of sound from record discs or magnetic recording tapes and feeds the signal to the amplifier.
- **Tone generator**—which produces the electrical signal equivalent of a tone, used for fire and other alarm signals, or produces the signal equivalent of church bells or chimes, used as input device with a sound system supplying loudspeakers mounted in a

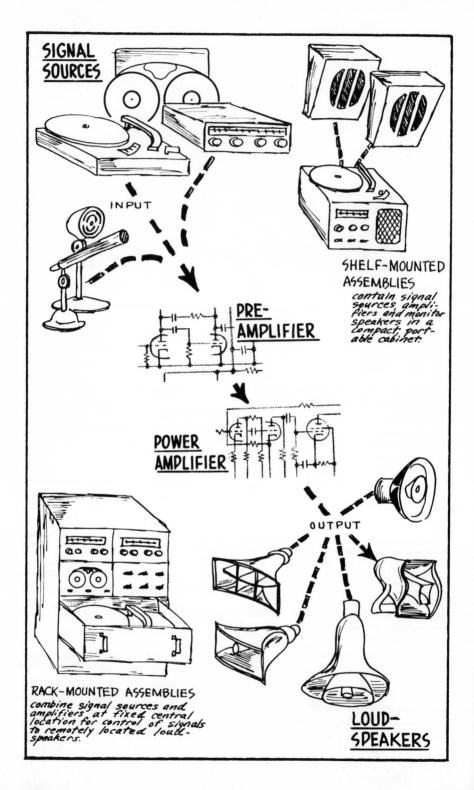

SIGNAL
SOURCES

INPUT

SHELF-MOUNTED
ASSEMBLIES
*contain signal
sources, ampli-
fiers and monitor
speakers in a
compact, port-
able cabinet.*

PRE-
AMPLIFIER

POWER
AMPLIFIER

OUTPUT

RACK-MOUNTED ASSEMBLIES
*combine signal sources and
amplifiers at fixed central
location for control of signals
to remotely located loud-
speakers.*

LOUD-
SPEAKERS

226

church steeple to simulate the sound of bells.

• **Musical instrument pickup**—which produces an electrical signal equivalent to the music being played on the instruments—such as an electric guitar.

• **Remote source**—which provides a signal input for the system over telephone wires, transmitted from a distant point outside the building in which the system is used.

• **Movie sound projector**—which produces the signal from the sound track of motion pictures.

Amplifier Units

From the source or input device, the signal is fed by appropriate cable connection to the system amplifier (one or more pieces of equipment). The amplifying equipment strengthens the signal and provides means for modifying the characteristics (bass, treble, etc.) of the sound which is produced by the system loudspeakers. The amplifying equipment generally consists of two separate sections: the preamplifier and the power or booster amplifier.

The preamplifier serves two purposes: it amplifies very weak signals—such as those from microphones, magnetic phonograph pickups and musical instrument pickups; and it permits combining the signal with other amplified weak signals or stronger signals like those from tuners, crystal phonograph pickups or tone generators. The preamplifier provides control of the relative volumes of the combined signals and provides control of the tone characteristics of the sound output.

From the preamplifier, the signal is passed along to the power amplifier—which may be on the same chassis or on one or more separate chassis. The power amplifier performs the function of greatly strengthening the signal to provide the necessary power to drive the loudspeakers connected to it.

In any system, the number and output ratings (in watts) of power amplifiers to be used depends upon the number of loudspeakers in the system and the level (in watts) to which each speaker must be driven to provide sufficient sound output for the particular application. For high output power requirements, a number of power amplifiers may be paralleled. In custom-assembled consoles or cabinet racks, the general practice is to install preamplifiers and power amplifiers relatively close to each other. In other installations, one or more preamplifiers may be mounted in a main operating center and connected by long lines to outlying power amplifiers located close to the loudspeakers they supply.

Although multi-chassis amplifier assemblies are commonly used in consoles and in cabinet racks for large custom installations, by far the largest number of sound installations employ an amplifier unit which combines both the preamplifier and power amplifier on a single chassis. Typical of the most widely used type of amplifier unit would be a single chassis unit with two input terminals for connecting microphones and an input terminal for connecting a radio tuner or record player. These terminals are commonly referred to as "inputs." Such an amplifier would be described as having "two microphone inputs and one phono input."

The amplifier would have an output of 30 watts, which is the amount of electrical power of the audio signal fed to the loudspeakers. This wattage rating should not be confused with the electrical watts consumed by the unit through the power cord which plugs into a 120-volt convenience receptacle. Such power consumption for a 30 watt amplifier might run around 140 watts.

Loudspeakers

The final element in the chain of components of a basic sound system is the loudspeaker. The amplified signal—the electrical equivalent of the sound which provided the signal in the source or input device—is delivered from the system amplifying equipment to the one or more loudspeakers in the system.

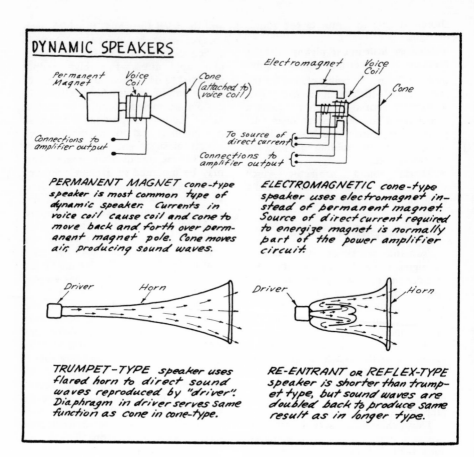

DYNAMIC SPEAKERS

PERMANENT MAGNET cone-type speaker is most common type of dynamic speaker. Currents in voice coil cause coil and cone to move back and forth over permanent magnet pole. Cone moves air, producing sound waves.

ELECTROMAGNETIC cone-type speaker uses electromagnet instead of permanent magnet. Source of direct current required to energize magnet is normally part of the power amplifier circuit.

TRUMPET-TYPE speaker uses flared horn to direct sound waves reproduced by "driver". Diaphragm in driver serves same function as cone in cone-type.

RE-ENTRANT OR REFLEX-TYPE speaker is shorter than trumpet type, but sound waves are doubled back to produce same result as in longer type.

The signal from the power amplifier output terminals is supplied to a loudspeaker over two conductors and applied to the voice coil of the speaker, usually through a very small transformer (about the size of a pack of cigarettes) called a line transformer.

Interaction between signal current in the voice coil and the magnetic field in the speaker magnet assembly produces motion of the cone or diaphragm of the speaker. The motion sets up sound waves corresponding to the alternations of the signal current which corresponds to the original sound waves. The loudspeakers therefore reproduce the original sound.

LAYOUT and design of modern sound systems for paging, public address or music distribution involves selection of basic components of proper size, type and capacity for the requirements of a particular installation.

A thorough understanding of the different types of equipment within the three basic equipment categories is essential to effective application of any sound system.

The following is a discussion of specific equipment types, starting with loudspeakers—the system output devices—and going through the system to the signal input devices.

Types of Loudspeakers

Basically, there are two types of loudspeakers used in paging and music systems:

Horn or trumpet type loudspeakers are capable of very high-power sound output. They have a very high efficiency of electric-power-to-sound conversion (over 15%) and are particularly suited to outdoor application—playgrounds, athletic fields, stadiums, etc.—and indoor applications where high sound power is required to cover large areas—auditoriums, factories, warehouses, etc.

Horn type speakers require no auxiliary enclosures or baffles and their rugged construction adds to their heavy duty, reliable, weather-resistant nature.

Horn speakers are made of metal in the form of a flared horn or trumpet. In this type of speaker, sound waves are set up by a small moving diaphragm in a magnet assembly called a "driver." This compact driver is coupled to the small end of the horn and directs the sound into the horn.

In passing through the horn, the sound waves are greatly strengthened in the same way that a megaphone strengthens the voice of a person shouting through it.

In this category will be found so-called reflex trumpets (for directional sound projection), radial reflex projectors (for uniform dispersion in all directions), explosion-proof speakers (designed for hazardous duty), submersion-proof speakers (which are immune to salt spray, gases, live steam, etc.), two-way and side-angle dispersion speakers (for covering corridors and large areas respectively).

The second basic type of loudspeaker, **the cone type,** consists of a sound producing diaphragm of pressed paper in the form of a shallow cone, driven by a voice coil and magnetic assembly similar to the one used for the horn speaker.

The cone speaker delivers sound directly to the air, without benefit of a horn. (It is possible to use cone speakers with wood or metal horn structures to increase their efficiency, but this is not generally done in commercial sound systems). The cone speaker is the type of speaker commonly used in radios, television sets and home hi-fi units.

Cone loudspeakers have a relatively wide-range frequency response, but their efficiency is much lower than that of horn type speakers.

They are preferred for use in music distribution systems wherever possible. And although they lack long distance penetration of trumpets and large area coverage from a single unit, their characteristics are ideal for paging and voice reproduction.

Cone type speakers are generally used in indoor applications where no severe acoustic problems exist. And even for outdoor applications, cone speakers are often used in special weatherproof enclosures.

For optimum results, cone type speakers should be used with a baffle or enclosure. Usually a wooden box or metal enclosure of some type, with the speaker mounted in it over a hole of proper size is used as the baffle.

There are two methods for properly connecting one or any number of speakers to an amplifier. The first is direct connection to amplifier output taps having an impedance value equal to the impedance of the single speaker or to the resultant impedance of the hookup of a number of speakers.

The second method for connecting loudspeakers to an amplifier involves the use of small transformers (about the size of a pack of cigarettes), called "line-matching" or simply "line" transformers.

The use of transformers permits mounting of speakers at great distances from the amplifier and provides the means for delivering varying amounts of power to individual speakers, to meet varying sound-loudness requirements.

Amplifiers

Most sound systems employ amplification units which combine the pre-amplifier and power amplifier on the same chassis as a complete unit. Complete amplifiers are available with

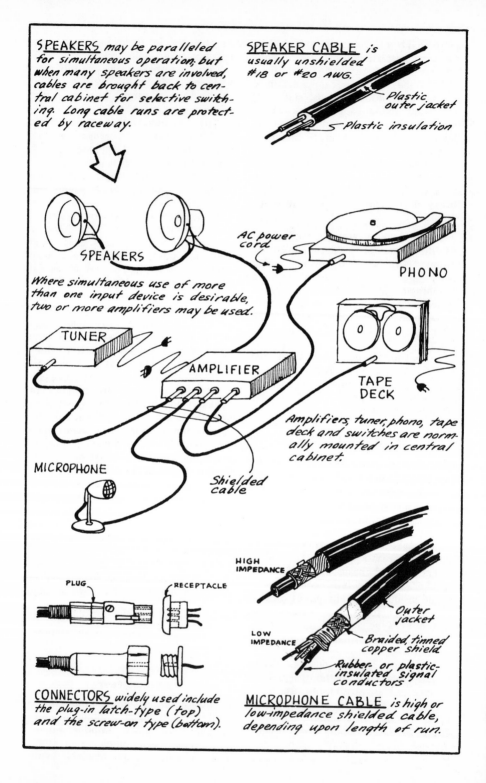

SPEAKERS may be paralleled for simultaneous operation; but when many speakers are involved, cables are brought back to central cabinet for selective switching. Long cable runs are protected by raceway.

SPEAKER CABLE is usually unshielded #18 or #20 AWG.

Plastic outer jacket

Plastic insulation

SPEAKERS

AC power cord

PHONO

Where simultaneous use of more than one input device is desirable, two or more amplifiers may be used.

TUNER

AMPLIFIER

TAPE DECK

Amplifiers, tuner, phono, tape deck and switches are normally mounted in central cabinet.

MICROPHONE

Shielded cable

PLUG

RECEPTACLE

HIGH IMPEDANCE

LOW IMPEDANCE

Outer jacket

Braided, tinned copper shield

Rubber or plastic-insulated signal conductors

CONNECTORS widely used include the plug-in latch-type (top) and the screw-on type (bottom).

MICROPHONE CABLE is high or low-impedance shielded cable, depending upon length of run.

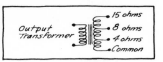
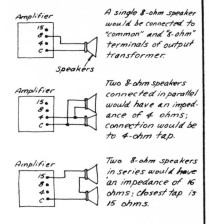
output running as high as 50 watts. This chassis will provide a variety of volume control knobs, tone control knobs, as well as microphone and phono input connectors and loudspeaker output terminals.

In sound systems running to 100 watts or more, separate preamplifier and power amplifiers are usually employed for greater flexibility. In installation of the latter type, one or more booster amplifiers may be used; and it is not uncommon to find large complex systems where a dozen or more booster amplifiers are connected to the output of a single preamplifier.

Packaged amplifiers for sound systems are available in many sizes and types. Most of the leading amplifier manufacturers provide engineering and assembly facilities in their factories to engineer complex assemblies of any size.

Complete amplification units incorporating a preamplifier and the power amplifier on a single chassis are readily available with the following output ratings: 10, 15, 30, 50 watts. Separate booster amplifiers are usually rated at 30, 50, 70, 100, 125, or 250 watts.

Where large amounts of output power are called for, any number of booster amplifiers may be incorporated in a single sound system and driven

from the preamplifier. Standard "packaged" amplifiers are available for systems requiring from one to five microphones and a phono or radio tuner.

The output impedance of the amplifier must be closely matched to the impedance of the speaker load if the rated output power of the amplifier is to be realized. Almost all p.a. amplifiers provide a relatively wide selection of output impedance taps.

A wide range of standard packaged amplifiers are available with phonographs fitted in their tops to permit using 33⅓, 45 and 78 rpm records up to 12-in in diameter.

MICROPHONES are commonly used as input devices for sound systems providing public address, paging and announcing.

Types of Microphones

Microphones may be classified according to their sensitivity pattern, the impedance of their outputs (high or low), principle of operation (crystal, dynamic or velocity).

• **By Sensitivity Pattern**—Microphones can be classified as uni-directional, bi-directional and omni-directional.

1. A uni-directional microphone

as might be expected, is sensitive to sound coming from one direction only and is selected frequently for stage and auditorium work. One of these is the cardioid microphone which enjoys tremendous popularity in night clubs and theater work because it rejects noise coming from the audience and provides pickup primarily from the performer or speaker. Its use is also indicated when acoustic feedback is a serious problem.

2. The bi-directional microphone will accept sound from both the front and back; it is, therefore, a logical choice for interviews, dialogue work, etc.

3. The omni-directional microphone does not discriminate against sound from any direction and is widely employed for group pickup, round-table discussions, etc.

• **By Output Impedance** (High or Low)—A high impedance microphone (which could be a crystal or dynamic type) must be operated within 50-ft of its amplifier. Also, high impedance microphone lines pick up hum and noise from nearby power devices, appliances and their associated power lines. If the distance between the microphone and the amplifier must be greater than 50-ft, a low impedance microphone, low impedance microphone cable and suitable amplifier should be employed.

Low impedance microphones may be operated over lines well in excess of 500-ft, without serious loss of signal level, and hum and noise pickup from adjacent appliances and power lines is considerably reduced.

• **By Principle of Operation**—There are three basic types of microphones: carbon, crystal and dynamic. The carbon microphone is almost never used in commercial sound systems. The crystal microphone is always a high impedance type, economical and capable of excellent performance.

The dynamic microphone is constructed like a miniature dynamic loud-speaker. Because of its ruggedness, wide frequency response, mod-

erate price and availability with either high or low impedance output, it is the most popular type in use today for commercial sound systems. Some high impedance dynamic microphones are assembled with a small switch at the rear of the instrument which permits the selection of high or low impedance.

Other Devices

Phono and tape mechanisms are other common input devices. Since the advent of the microgroove record, it has been possible with the automatic record changer to provide a program of music distribution throughout a factory which runs to several hours without requiring attention. Tape play-back mechanisms are available today with specially prepared tapes to provide suitable musical programming for periods up to eight hours.

Radio tuners differ from AM and AM/FM radios in that the audio amplifier section and loudspeaker have been omitted. The radio tuner permits selection of broadcast programs which may be introduced into the amplifier through the phono or tuner input.

Tone generators are commonly used in the larger console and vertical cabinet rack assemblies installed in factories. A typical tone oscillator may be easily connected to the time clock in the factory so that a steady tone of short duration will be distributed to the loudspeakers throughout the factory at regular intervals to indicate start and termination of lunch periods, work shifts, etc.

Hardware

Special types of wires, cables and connectors are used for interconnecting the various elements of sound systems. Conductors and connectors commonly used for power and light wiring are unsatisfactory for use with sound systems and could result in troublesome operation.

Important characteristics of sound

systems hardware are as follows: Microphone cables are special cable assemblies used to provide proper connection of microphones to amplifier input terminals. There are two types of microphone cable—high impedance cable and low impedance cable.

High impedance microphone cable is used with high impedance microphones, where the length of the connection between the microphone and the amplifier does not exceed 50 feet. This cable consists of a single rubber (or plastic) insulated stranded conductor enclosed in a braided (or spiral) tinned copper sleeve. An outer sheath of rubber or plastic protects and insulates the complete assembly. The single conductor carries the signal, and the shield is used for a return conductor, connected to the ground at the amplifier. This type of cable—because it is used in short runs—is generally run exposed and not pulled in conduit.

Low impedance microphone cable is used with low impedance microphones, which are required where the microphone is located more than 50 feet from the amplifier. This type of cable consists of two insulated conductors within an overall copper braided sleeve and a protective sheath. The two wires carry the signal and are electrically balanced to ground which is provided by the shield. A low impedance input line must be terminated in a low impedance input terminal on the amplifier. Since such lines are commonly used for long runs—over 50 feet, up to 500 feet—they are generally pulled in conduit to provide sufficient mechanical protection and proper installation.

Phono leads for connecting record players to amplifiers are invariably single, shielded conductors—the high impedance type described above.

Cables run from the output taps on an amplifier are called "output leads." These are usually unshielded plastic or rubber insulated conductors made up in the form of twisted pairs of wires in an overall outer jacket.

Special connectors are made for low impedance cable and for high impedance cable and for phono and speaker leads.

Closed Circuit TV

ONE OF THE MOST interesting applications of communications equipment in modern industry and commerce is closed circuit television. Available as plug-together systems of basic component units, closed circuit television or industrial television, as it is also described, is a method of remote viewing. In industry, research, commerce, education and science, such systems provide continuous observation of objects, operations or conditions which for reasons of danger, distance, inaccessibility, inconvenience or cost cannot be viewed directly.

System Operation

Typical closed circuit television systems consist of three basic pieces of equipment: the camera, the control unit and the monitor or receiver. Each is used as follows:

The camera unit is about the size of a small table model radio. It has an external lens assembly which is used to focus on the scene to be

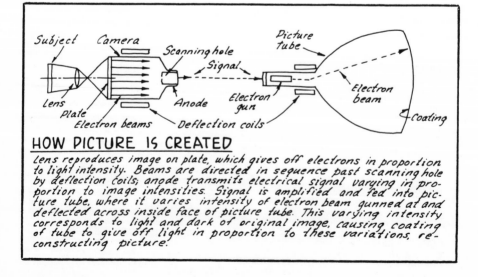

HOW PICTURE IS CREATED

Lens reproduces image on plate, which gives off electrons in proportion to light intensity. Beams are directed in sequence past scanning hole by deflection coils; anode transmits electrical signal varying in proportion to image intensities. Signal is amplified and fed into picture tube, where it varies intensity of electron beam gunned at and deflected across inside face of picture tube. This varying intensity corresponds to light and dark of original image, causing coating of tube to give off light in proportion to these variations, reconstructing picture.

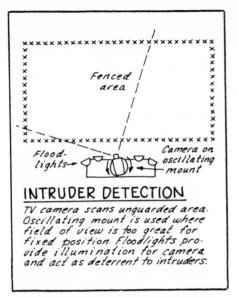

INTRUDER DETECTION

TV camera scans unguarded area. Oscillating mount is used where field of view is too great for fixed position. Floodlights provide illumination for camera and act as deterrent to intruders.

there is real need for detail in the image to be electrically transmitted.

Another variation in cameras is made in the lens systems. Where the distance from the camera in its mounted location to the scene which it is recording is fixed, the focus of the lens assembly can be set and left. But in other cases where the camera in a fixed location is picking up images of objects which are moving with respect to the camera, there must be some provision for adjusting the focus of the lens.

This requires motor actuation of the lens assembly, with remote control of the focusing from the point of the receiver where the picture is being viewed. And for some applications, the camera may require a remote controlled turret on the camera to provide changing of lenses, such as for wide angle or telephoto viewing.

In typical applications, the camera or cameras used in a system are fixed in location where they continuously "photograph" the scene which is to be observed from a different location. The electrical signal output from the camera is fed over special wires to the system's control unit, which may be mounted close to the camera unit or may be as much as 1,000 feet away.

transmitted. A special tube within the housing converts the image from the lens into an electrical signal which it sends to the control unit.

Camera units for various systems differ somewhat in size and picture-taking capability. Some cameras are much more sensitive than others, requiring only very low levels of light for acceptable pickup of the image. Other cameras require very high lighting on the scene to be transmitted. Cameras also vary in ability to "see" detail. In many cases, such as observation of a static object like boiler gauges, there is no need for fine detail in the image. But in other cases, such as using the camera for viewing license plates or the faces of people seeking entrance to a plant,

The control unit in a typical system is supplied with the signal which is the electrical equivalent of the image produced in the camera unit. This unit contains vacuum tube circuits which control operation of the cam-

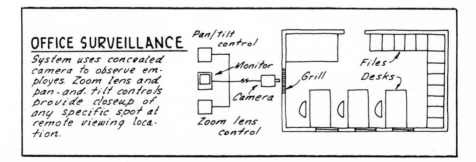

OFFICE SURVEILLANCE

System uses concealed camera to observe employes. Zoom lens and pan-and-tilt controls provide closeup of any specific spot at remote viewing location.

Pan/tilt control

Monitor

Camera

Zoom lens control

Grill

Files

Desks

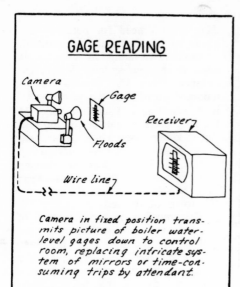

GAGE READING

Camera

Gage

Receiver

Floods

Wire line

Camera in fixed position transmits picture of boiler water-level gages down to control room, replacing intricate system of mirrors or time-consuming trips by attendant.

era and coordinate its output with operation of the one or more system receivers.

In typical systems, the control unit may be located as much as 1500 feet from the receiver, or up to 4000 feet in some cases with the use of line amplifiers. Inexpensive coaxial cable is used to connect the control unit to the one or more system receivers located at points where observation is required.

Power supply for operation of typical systems consists of simple plug connection to 120-volt, single-phase ac receptacle. Power consumption usually runs around 200-250 watts. Electric supply to scene illumination is extra.

The monitor in a closed circuit television system is very similar in appearance to the standard home television receiver. It contains a cathode ray tube which provides the screen on which the transmitted scene is recorded. Although some systems use special monitor units with characteristics designed for the system itself, some systems can use both their own TV receivers. Monitor screens vary from 5 to 14 inches, and larger.

Closed circuit television systems

also include a number of accessory equipments: remote control zoom lenses for cameras, acoustical and explosion proof housings, portable monitors, remote control pan and tilt mounts for the camera and weatherproof housings with windshield wipers.

Applications

Common uses of closed circuit systems are as follows:

1. For watching smoke stacks from inside a building, with the camera located away from the building to view the stacks.

2. For transmitting school instruction from one point to a number of rooms within the school building.

3. For observing distant signals and operations from the inside of a railroad yard control tower.

4. For watching the gates of industrial plants from within the plants, or similar guard applications in security areas.

5. For viewing the internal operation of giant power boilers where combustion is too hot and too brilliant for the naked eye.

6. For close-up viewing of the pouring of molten steel from the operator's position 60 feet away.

7. For close-up viewing of meters and other equipment in hazardous testing operations where the test operators must be located remote from the actual test.

8. For viewing of hospital rooms from the nurse's duty station on each floor of hospitals.

9. For transmitting a continuous picture of gauges high up on boilers in electric generating plants, to provide viewing of the gauges from the operator's control station.

10. For use in hospitals to permit medical students in one room to view the details of operations going on in the operating room.

11. For many industrial processes to provide viewing of distant controlled operations.

Most closed circuit TV manufacturers provide system design service.

Industrial Electronics

THE field of industrial electronics embraces an extremely wide and varied assortment of equipment, component parts and supplies. This includes the many industrial and commercial control and power devices which utilize electronic circuitry in their operation—from motor controls and welding controls to data processing machines, radio and TV broadcasting equipment and laboratory power supplies—to the almost endless inventory of parts which go into their manufacture and maintenance.

The term "electronic" is commonly applied to equipment which contains electron tubes or transistors and their associated circuitry. This covers devices in which the design is based on the flow of current in electron form, usually through vacuum or gas-filled tubes or through solid devices known as semi-conductors. Although this type of current flow can be high power and high amperage in large rectifiers for dc energy supplies, electron current flow is frequently in thousandths of an ampere (called milliamperes) or even millionths (called microamperes). For this reason, it is commonly said that the difference between electricity and electronics is the placement of the decimal point, e.g., 100 amps at 120 volts as against 0.001 amps at, say, 12 volts.

Logically, the term electronics when applied to components has been expanded to include the parts peculiar to use in electronic circuits—resistors, capacitors, relays, coils, small transformers, rheostats, chokes, connectors, terminals, small switches, test equipment and accessories.

Electronic Equipment

Modern equipment and apparatus which is electronic in nature includes:

Motor Control—A number of types of electronic motor controllers are commonly used in industrial electrical systems. All such devices are first of all motor controllers—that is, they are identified according to their function. To arbitrarily call such devices "electronic controllers" places prime emphasis on the nature of their operation rather than their function and tends to suggest a line of demarcation between such devices and other motor controllers which are not electronic in the nature of their operation. In fact, many applications can very well use electronic or non-electronic controllers, provided they have similar functions. Many electronic control applications represent only more efficient ways of doing what was formerly done by non-electronic devices. With motor controllers and with other control and power equipment, the term "electronic" should not be used when concern is with the function of the equipment. Such usage only adds confusion and is often cause for rejection of the equipment by the customer who equates the term with undue complexity.

Electron tube circuits are often used in control of electric motors and generators. Such uses include control of motor speed and control of generator output voltage. A dc shunt motor, for instance, may be varied in speed by varying the voltage applied to its armature or the voltage applied to its shunt field. Such speed control can be accomplished by use of variable resistance in the armature circuit and in the shunt field circuit, with dc power applied to the motor leads. Such application is non-electronic.

Another method of speed control for the same motor might use a dc generator driven by an ac motor to supply a variable voltage to the armature, with a smaller dc generator to supply a variable voltage to the shunt field. This method is still non-electronic, but it has certain advantages over the first method where the load on the controlled dc motor varies.

A third method of accomplishing the same results makes use of electronic tube rectifier circuits to replace the bulky, expensive and relatively inefficient ac-dc motor-generators. In such a case, the controller function remains the same but the nature of operation is different. And today, the electron tubes in such circuits, thyratrons or ignitrons, are being replaced by much smaller, more efficient solid-state devices known as silicon rectifiers — from the family of devices known as semi-conductors. In any such circuit, the number of rectifier units and the actual circuitry will vary with the size of the motor. In smaller size units, selenium rectifiers —solid-state discs—are often used, or even vacuum-type amplifier tubes are used. And along with these devices goes a variety of circuit parts.

Other electronic circuits are incorporated in motor control equipment to provide automatic regulation of motor speed. This may involve changes in speed to meet requirements of the driven load or it may require maintaining a constant speed output from the motor. Of course, manual control of speed is provided by the types of controllers described above. In such cases, the operator uses his brain and hand to correct for any errors in speed—adjusting the speed as required. In automatic regulators, the circuitry itself constantly checks the actual motor speed against an input signal which represents the desired speed and makes changes in speed when necessary.

Many industrial processes and operations depend upon effective motor speed control. A typical example is any form of material reeling. Say cloth or paper is reeled by being fed to a motor-driven drum. With the speed of material feed to the drum constant, the driving motor must reduce its speed as material is added on the reel, making its diameter larger and taking more material from the feed on each successive revolution of the drum. An automatic speed regulator for the motor could use an input signal based on the tension in the material in the feed to adjust the speed for constant tension In this way, the speed will change as required by increasing diameter of the reel.

Electronic speed controllers and voltage regulators for use with industrial systems are made in compact enclosures or panelboard-type layouts for ready installation and relatively simple connection into their asso-

HALF-WAVE RECTIFIER

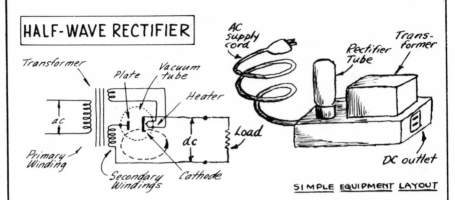

SIMPLE EQUIPMENT LAYOUT

Simplest electronic rectifier produces dc pulses across **output terminals** which may be used to feed a dc device. Transformer is used to reduce or increase ac line voltage to the value required for external dc device, and to provide low voltage for tube heater.

Since ac voltage is constantly reversing, plate of tube is alternately positive and negative. When the plate is positive, electrons flow from the hot cathode to the plate, through the transformer and load resistance and back to cathode. When plate is negative, no current flows. Thus current **through load** is always in same direction, although it varies between a minimum and maximum value each time the tube conducts.

FULL-WAVE RECTIFIER

The addition of another tube eliminates the "on-off" operation. When Tube "A" is not conducting, tube "B" is. A study of the circuit will show that current through the load resistance is always in the same direction. (Heater and heater winding have been omitted from diagram for simplicity.)

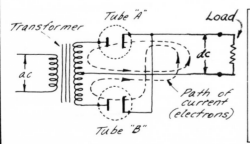

the ELECTRONIC RECTIFIER

is a fundamental part of much electronic equipment. An understanding of the basics shown here will be helpful in studying more complicated circuits.

ciated drives. Such equipment is completely packaged and may be standard off-the-shelf models or custom-built controllers. Units are available which provide for ready connection of ac power to the controller input and connection of the dc motor to the output terminals. Even complete, automatic control, variable-speed drives are packaged for line connection to an ac circuit and load connection to the shaft of the dc drive.

ANOTHER category of control equipment which makes effective use of electronic circuitry covers controllers for resistance welders. Basically, a resistance welder is a device used to fuse or weld two pieces of metal together by passing a large current (in the thousands of amperes) through the junction between the two pieces while they are held tightly together.

The welding machine contains a special transformer which receives energy at the distribution voltage (say 480 volts) and relatively low current and transforms it into energy at much lower voltage but at the very high current required for welding. A special form of controller is used to effect proper operation of the machine for use.

The basic function of a welding controller is to close and open the circuit to the welding transformer for the split second it takes to make a weld. Speed in switching ON and OFF is therefore an important requirement in a welder controller. If the controller can turn the machine ON, hold it ON for only the part of second necessary and then turn it OFF instantaneously, no power will be wasted and the machine will be able to make the maximum number of weld operations in any period of time. Welders are frequently required to make hundreds of welds per hour.

To accomplish the above control function, ordinary magnetic contactors are commonly used to do the

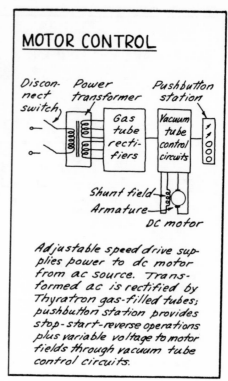

MOTOR CONTROL

Discon-nect switch; Power transformer; Pushbutton station

Gas tube rectifiers — Vacuum tube control circuits

Shunt field — Armature — DC motor

Adjustable speed drive supplies power to dc motor from ac source. Transformed ac is rectified by Thyratron gas-filled tubes; pushbutton station provides stop-start-reverse operations plus variable voltage to motor fields through vacuum tube control circuits.

ON-OFF switching. But, the inertia of their operation—the relative slowness of the heavy moving parts—limits the precision of timed current application and reduces the repetitive rate of welding operations. And the nature of magnetic contactor operation imposes a considerable maintenance task and makes a lot of noise.

As an alternative to the magnetic contactor, electronic-type contactors are commonly used to control resistance welders. Such controllers make use of the switching capabilities of certain types of electron tubes. A common tube used in such controllers is called the **ignitron** tube. This is a special type of heavy-current rectifier tube which is used in pairs to provide ON-OFF switching of the ac supply to a welder transformer. The ignitrons and their associated circuitry are mounted in a suitable enclosure for ready installation at welder locations. The operating char-

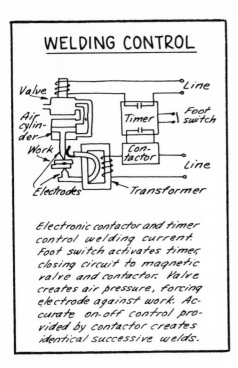

WELDING CONTROL

Electronic contactor and timer control welding current. Foot switch activates timer, closing circuit to magnetic valve and contactor. Valve creates air pressure, forcing electrode against work. Accurate on-off control provided by contactor creates identical successive welds.

acteristic of this type controller makes it much faster (there are no moving parts) and more precise, providing better welds in shorter time.

Ignitron welder controllers are generally rated in amperes, to designate their switching capability in the same way that a magnetic contactor is rated. They are commonly made for use on 220-volt circuits and 440-volt circuits, with accessory control equipment which also makes use of electron tube circuits. These auxiliary circuits may use a number of different types of tubes to provide control of the time and operation of the moving electrodes on the welders, for automatic operation.

Packaged DC Supplies

Rectifiers are elements of electronic circuitry used to convert alternating current into direct current. The term rectifier is used to describe either the converter element itself—a tube or semi-conductor device—or the complete packaged assembly which is designed for ac input and dc output, with necessary control devices. Considering the packaged type rectifiers, there are sizes ranging from small low wattage units to kilowatt units. The small units are made with either high-current, low-voltage output for plating or battery-charging applications or with voltages running to several hundreds and low to moderate current outputs for operating dc motors or appliances or for powering other electronic equipments for control, signals, communications or test purposes.

The larger sizes of packaged rectifier units, for producing more than one or two kilowatts of dc power, are commonly designated for connection to three-phase ac supplies and may use many rectifier tubes or semiconductor devices to obtain the required current output. And in these rectifier units—just as in the smaller units—transformers are used to step-up or step-down the ac voltage supplied to the unit, to obtain the required dc output from the rectifier.

Although gas- or vapor-filled tubes were for many, many years the standard rectifying elements in large and small dc power supplies, developments in recent years in the field of solid-state devices have made such devices the prime elements for rectification from ac to dc. Selenium disc rectifiers are made in packaged units and find very effective application where high currents at low voltage are needed. For dc power needs at hundreds of volts, germanium and silicon rectifier elements provide almost 100% efficiency of conversion. Typical power rectifiers in the latter class are rated up to 25 and 40 kw, and even higher.

Time-Delay Relays

Other control devices which use electronic circuits for their operation include time-delay relays. Again there are many good and effective ways of achieving time-delay in electric circuits—i.e., ways of operating a contact at a definite time after a signal

PHOTOTUBE

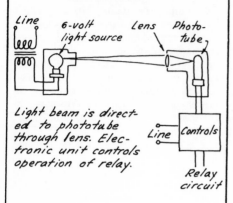

Line   6-volt   Lens   Photo-
        light source   tube

Light beam is direct-
ed to phototube
through lens. Elec-   Line Controls
tronic unit controls
operation of relay.
                        Relay
                        circuit

SIMPLE CONTROL CIRCUIT:

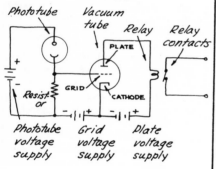

Phototube   Vacuum
            tube   Relay   Relay
                    contacts
                PLATE

+

−   GRID
  Resist-   CATHODE
    or
        −||+ −||+

Phototube   Grid   Plate
voltage   voltage voltage
supply   supply supply

With no light on phototube no current flows through tube. Grid of
vacuum tube is kept negative
by grid voltage supply; hence
no current flows through vacuum tube, and relay contacts are
closed.

When light strikes phototube,
current flows upward through
tube, making top of resistor and
grid less negative. Current flows
upward through vacuum tube,
operating relay. A beam of light
on the phototube will keep contacts open. If beam is interrupted, contacts will close.

is given. Standard dashpot relays utilize the retarding action of a piston pushing into a hydraulic cylinder with a hole at the end to provide controlled slow movement of the piston and consequent delayed closing or opening of contacts with which the piston is mechanically associated. Or a bellows with controlled air escape can provide the same action. But again, electronic circuitry offers a superior type of time-delay action for certain applications. And as with the welder controller, elimination of some moving mechanical parts improves the performance and reduces the maintenance needs. Such time delay relays use tubes to supply dc current to the coils of relays, with resistor and capacitor circuits to control the delay in operation of the tubes. Extensive use is made of this type of relay in many more complex electronic control circuits.

Modern time-delay relays are mounted in small enclosures, with adjustable settings for time-delay action. Such relays are widely used in conjunction with motor controllers and various industrial process operations.

A POPULAR CATEGORY of electronic control is made up of the various devices used to control theatre and display lighting and devices used to regulate electric heaters for industrial furnaces. In such equipment, tube circuits have been used to indirectly vary the ac load current to the lamps or heaters. The load current does not pass through the electronic circuitry. Instead, the tube circuits regulate a control current which flows in a magnetic device known as a saturable reactor.

The use of saturable reactor control of alternating current has many advantages over non-electronic type of control. For instance, ac current through a group of incandescent lamps can be varied very simply by connecting a rheostat or variable resistor in series with the lamp load. By setting the rheostat for lower or higher values of resistance, more or less current will flow through the load, thereby controlling the amount of light output from the lamps. Although such control is satisfactory for relatively small values of current, rheostats for large current values have such large size, difficult operation and substantial heat losses that they are economically and practically undesirable.

A saturable reactor connected in series with the above lamp load offers just as effective dimming control with much smaller size equipment and much lower losses—even for heavy ac load currents. The basic operation of the saturable reactor is as follows: **1.** The reactor consists of a single main winding on a laminated steel core (like a transformer, but only one winding), connected in series in the line conductors to the lamp load. **2.** The reactor has a smaller control winding on the core. **3.** When no current flows in the control winding, the main winding connected in series with the load has a high value of inductive reactance which limits the amount of alternating current flowing to the lamps. **4.** But when direct-current is fed to the control winding, the value of the main winding inductive reactance is reduced in proportion to the value of dc current and there is less opposition to the flow of the ac load current, causing the lamps to glow brighter. Thus the value of load current can be controlled from practically "OFF" to full "ON" by varying the comparatively small value of dc control current. And this basic action of the saturable reactor is applied with modifications in various types of control devices.

In typical theatre dimmer systems, the reactors and tube control circuitry are placed in a convenient out-of-the-way place, with one or more light-adjusting levers at the control station near the stage. The tube circuits through which the control windings are energized use vacuum and gas-filled· tubes—thyratrons and phanotrons are typical—and usual electronic parts like resistors, capacitors, small transformers, relays, etc. And

SATURABLE REACTOR

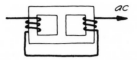

Wire wound on two legs of a core as shown forms an inductor, or reactor. Inductive reactance formed by the resulting magnetic field tends to reduce the current flowing in the wire.

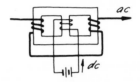

A saturable reactor is formed if another coil is wound on the center leg of the core and dc current is passed through it. The inductance presented by the ac coils is reduced, increasing the ac current.

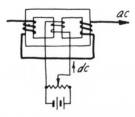

Adding a variable resistor in the dc circuit permits the dc current to be varied; thus the ac current can be increased or decreased as desired. By replacing the battery with some type of electronic rectifier, the need for a separate dc source is eliminated.

LAMP DIMMING

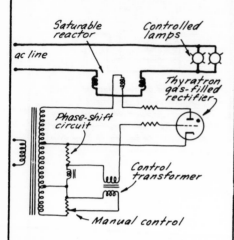

Saturable reactor

Controlled lamps

ac line

Thyratron gas-filled rectifier

Phase-shift circuit

Control transformer

Manual control

In this application of a saturable reactor the thyratron is used as a rectifier to provide dc for the center coil of the reactor; lamp current flows through the ac coils. By operation of the manual control, the dc current varies, and the lamp current is increased or decreased. The phase-shift circuit is used to produce a time lag between thyratron grid and plate voltage, making a variation in current possible through the tube.

MAGNETIC AMPLIFIER

A magnetic amplifier is essentially a saturable reactor with an associated rectifying device. It is used to control larger power outputs by small power inputs. Its advantages include:
- *No moving parts*
- *Sturdy construction*
- *Good efficiency*
- *No warm-up time required*
- *No dc line necessary*
- *Isolates input and output*

equipment used for controlling heat output (temperature) of electric furnace heaters is generally similar in the use of the reactor and its control, with a little more involved input control to provide regulation of temperature.

Higher Frequencies

Although all of the foregoing types of industrial electronic equipment—motor controls, welding controls, rectifier packs, relays and lighting and heating controls—operate with direct current and standard 60-cps alternating current, the vast majority of electronic devices in use today operate at much higher frequencies. This includes many types of industrial equipment and the wide range of communication equipment. Such equipment is infinitely more complex from the standpoint of electronic circuitry. And unlike the previously described industrial electronic equipment, many high frequency devices cannot be duplicated in their functions by non-electronic devices. It is the electronic nature of such devices which makes possible operating frequencies of thousands or millions of cycles per second.

Some typical high-frequency electronic applications which fall in the category of industrial electronics are as follows:

High-frequency heating—This covers induction and dielectric heating, which is used for various industrial work heating tasks. Electric heat offers speed, ease, cleanliness and versatile control as compared to conventional fuel-fired heating. Induction heating involves the use of a coil of wire through which high-frequency current is passed. Any conducting material can be heated to almost any temperature by placing the material within the coil of wire. Rapidly reversing currents are induced in the material, **producing** heat by their presence. These are called "Eddy" currents. In magnetic material, like iron or steel, hysteresis—the inertia of magnetically oriented particles—also contributes to the heating.

Electronic circuits are used to generate and control the high-frequency current supplied to the induction coil. Typical frequencies for electronic equipment used with induction heating equipment range from about 10,000-cps up to about 1,000,000-cps.

Dielectric heating is an electronically powered industrial heating technique applicable to materials and parts which are not conductors of electrical current. Such equipment usually operates at frequencies from one megacycle up to about 50 megacycles (abbreviated "mc"). The work to be heated is placed between metal plates. The output leads from an electronic oscillator (the high-frequency generator) are connected to the plates. This produces rapidly-reversing dielectric stresses in the work, causing strong molecular agitation which heats the work. A common use for dielectric heating is to dry glue between the wood layers in plywood.

Ultrasonics—This is an electronic application in which high-frequency vibrations are used to produce agitation in solid materials. The frequency is generally above 20,000-cps (the common upper limit of human audibility, hence the term "ultrasonics"), up to 200,000-cps. Ultrasonic vibrations are used for washing clothes or dishes or other solid objects by applying the vibrations to the water in which the objects are immersed. The vibrations in the water literally shake or scrape apart foreign particles on the objects. Such vibrations are also used for mixing liquids together, for measuring distance through water and for certain cutting operations. Electronic circuitry produces the required current alternations which are then applied to special types of crystals or ceramics which convert the electrical alternations to physical vibrations.

A LTHOUGH MANY applications have been found for electronic circuitry in industrial power and control equipment, the most widespread use of electronics has been in the field of communications.

Transmitting stations for broadcast radio and television consist of relatively large and complex equipment installations. As such, they represent substantial markets for all types of electronic assemblies and supplies for maintenance.

Transmitters for commercial radio systems are small-scale versions of broadcast stations. Completely packaged transmitting units and receivers are available for such application and require very simple installation.

Two-way radio for wireless communication between two or more points, made in a variety of packaged assemblies for use in cars, trucks, boats or other mobile applications, consists of transmitting and receiving units in single-assembly housings.

Closed-circuit television is available in complete systems of packaged units which require only interconnecting wires for operation. Typical closed-circuit TV (or ITV—industrial television, as it is frequently called) is used in industry for various remote viewing applications. It may be used for observing a dangerous process by putting the pickup camera at the dangerous point in the process and putting the receiver or viewing monitor at some remote, safe point from which the process can be watched. Many similar remote viewing uses are possible.

Sound systems are also made up from completely packaged and readily interconnected units—microphones, record and tape players, amplifiers and loudspeakers. Such systems provide for public address, paging, music distribution in industrial plants and can be made for two-way wired communication.

Intercom is a wired system of two-way or multi-way communications. Such systems are basically made up of microphones, amplifiers and loudspeakers with suitable controls. Wiring between units can provide for communication among any number of points.

Electronic Parts

Electronic equipment—both industrial power and control devices and the complete range of communication units—is constructed of basic parts which are grouped in a category known as "electronic parts and supplies." These include an almost endless variety of little components which are sold to original equipment manufacturers for use in production line construction of various electronic assemblies, to industrial plants for use in the repair and maintenance of their electronic equipment, to radio and television stations, to other radio transmission facilities, to radio and TV service organizations and to other technicians involved with electronic circuitry.

Tubes—This is the basic category and involves very large quantities of expensive items. There are several classifications of tubes based on application. Tubes for use in radios, television sets, sound systems and other packaged communication sets are referred to as "receiving tubes." Another classification is that of "special purpose tubes" and covers the many tubes used in industrial electronic equipment and in radio, television and commercial transmitting facilities. Tubes in this class are generally more expensive than the receiving type tubes. Some such tubes sell for $200, $400 and even $700 apiece. A third class of tube is the cathode ray tube which includes tubes for oscilloscopes and television picture tubes.

Transistors—This is a growing category of components. These tiny devices are finding ever wider application as substitutes for vacuum and gas-filled tubes. They are capable of performing many of the functions of tubes with great benefits in the way of reduced heat output and longer life. As with vacuum tubes, there are both voltage amplifier and power amplifier types of transistors.

Rectifiers—In addition to tube type rectifiers, which are included in the types described above, there are many so-called dry-type rectifiers made of semiconductors such as selenium, germanium, or silicon which can duplicate the functions of tubes again with appreciable advantage, not the least of which is very high efficiency. A special category of dry-type rectifier is the diode.

Resistors—These units vary widely in size, ratings, construction and application; but they are bread-and-butter components in electronic circuits. The most common type of electronic circuit resistor is a little composition cylinder about a half-inch long and a quarter-inch in diameter, with a bare lead coming out of each end. Other types do not look like this at all, but they function the same.

Capacitors—Another basic item, capacitors, are small units. Some are mica or ceramic, about the size of a dime; some are molded plastic or paper tube, varying from the size of a cigarette butt to that of a cigar butt; and others—the electrolytic type —are paper or metal cylinders up to about five inches long and about two inches in diameter. Still another type of capacitor is the oil-filled type widely used in industrial gear. This is generally in a metal can of rectangular cross-section. In electronic application the capacitor is commonly called a "condenser."

Transformers and coils—A special line of power and signal transformers is made for use in electronic equipment. Power units are used for stepping ac supply voltages up or down

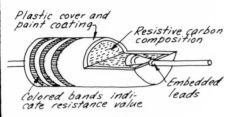

Plastic cover and paint coating

Resistive carbon composition

Colored bands indicate resistance value.

Embedded leads

FIXED CARBON RESISTOR,

partly cut away to show construction. Variations in carbon mixture and concentration provide a wide range of resistances.

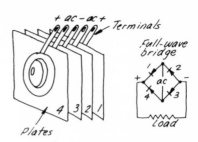

+ ac − ac + Terminals

Full-wave bridge

Plates

METALLIC RECTIFIERS

convert ac to dc without moving parts, due to difference in resistance to current flow, depending upon direction of current. Typical junction materials used include copper oxide, selenium, silicon, and magnesium copper sulfide.

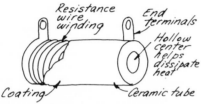

Resistance wire winding

End terminals

Hollow center helps dissipate heat

Coating

Ceramic tube

FIXED WIREWOUND RESISTOR,

generally used as a load resistor, handles heavy currents. Resistance wires may be left uncovered or coated with cement, lacquer, or vitreous enamel.

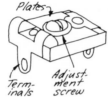

Plates

Adjustment screw

Terminals

TRIMMER CAPACITORS

are used to provide small adjustments in circuit capacitance.

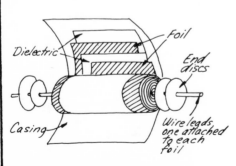

Foil

Dielectric

End discs

Casing

Wire leads, one attached to each foil

PAPER CAPACITOR consists of

rolled metal foil electrodes separated by impregnated paper dielectric. An improved type uses a metallized coating on one side of paper.

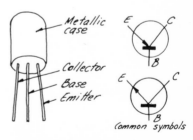

Metallic case

Collector

Base

Emitter

Common symbols

TRANSISTORS perform many

of the functions of vacuum tubes. Their operation depends upon the transfer of electrons at the junction of component semiconductors when a voltage is impressed across them.

prior to feeding a rectifier circuit from which dc operating voltages are obtained. Signal transformers may be iron-core devices—such as audio output transformers or microphone input transformers—or they may be air-core units for use in radio-frequency circuits. Various coils used in electronic circuits may also be iron-core devices—such as filter inductors or chokes or high-frequency coils with movable powdered iron slugs for varying the inductance.

In addition to these types of devices, there are special wires and cables for electronic work, special controls which incorporate the characteristics of resistors or capacitors or inductors, an infinite assortment of hardware for assembly of electronic chassis, and special tools, supplies and instruments.

Tools

THE VARIETY of tools used in electrical construction and maintenance work is almost unlimited. Such tools range from small hand tools to very large power driven equipment. The following is a rundown on typical tools and their applications.

Hand Tools

Of course, the most common tools in this category are the various types of pliers, wire cutters, screw drivers, wrenches and penknives used by electricians.

These and other small hand tools need no elaboration. There are, however, a number of special hand tools:

Wire Strippers are made in a number of hand types for ready skinning of insulation from wires for splicing and terminal connections, without nicking the conductor or cutting any strands.

Fish Tapes are made in several types to provide ready passage through conduit and duct to pull circuit conductors back through the raceway. The steel fish tape is the old standby, consisting of a length of flat or round steel for pushing and working through raceway. One such fish wire comes with a reel which serves as a handle for the mechanic working the tape, providing positive grip and tape control.

Another fish line device consists of a hand gun which uses a jet cartridge to "shoot" a nylon cord through the raceway. Still other compressed-air methods are used for fishing through conduit.

Label Embosser is a hand-held pliers-like device for embossing metal or plastic tape for use as identification labels on conductors, terminals, cabinets, etc. The device contains adjustments for the numbers and letters used to make the labels. The embossed tape is readily attached by its pressure sensitive back.

Conduit Tools

Cutting Tools for conduit range from the hand hack saw to the speedy metal cutting band saw which can be used to cut bundles of conduit. High-speed, power driven, abrasive cut-off wheels are also used.

Conduit Threaders are today primarily power driven machines which do a highly efficient cutting and threading job for all sizes of conduit. Priced well within the tool budget of any contractor, these units are readily transported on the job. For threading at-point-of installation, power threaders are made in hand-held, portable types. Such devices can be used to operate dies and conventional thread-

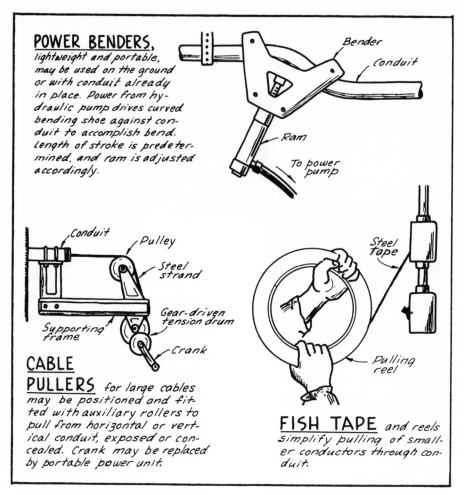

POWER BENDERS, *lightweight and portable, may be used on the ground or with conduit already in place. Power from hydraulic pump drives curved bending shoe against conduit to accomplish bend. Length of stroke is predetermined, and ram is adjusted accordingly.*

Bender

Conduit

Ram

To power pump

Conduit

Pulley

Steel strand

Gear-driven tension drum

Supporting frame

Crank

CABLE PULLERS *for large cables may be positioned and fitted with auxiliary rollers to pull from horizontal or vertical conduit, exposed or concealed. Crank may be replaced by portable power unit.*

Steel Tape

Pulling reel

FISH TAPE *and reels simplify pulling of smaller conductors through conduit.*

ers for up to 6-in conduit. They can also be used as cable-pulling winches and earth augers.

Conventional cutting and threading machines are made in floor-standing models with integral electric drives and complete accessories for the modern conduit job.

Conduit Benders are made in both manual and power-driven models to cover the full range of sizes of rigid steel and aluminum conduit and thin-wall (EMT) conduit. For the smaller sizes (up to 1 " conduit) of conduit and tubing, the hand "hickey" is considered the most practical and efficient tool. For all type of bends and off-

sets in the smaller conduit sizes, hand benders on portable pedestals provide accurate control, are fast loading and easy to use. One such type is the ratchet bender which permits short strokes to make the required bend.

Power benders using hydraulic pistons powered by hand-operated or electric driven pumps are made in many modern types for quick and easy bending of large size conduit. Such benders are made in table mounted units or completely portable assemblies of bending-shoe-and-piston unit with separate hand pump. The latter unit can be used to bend and offset conduit in any position, even conduit mounted overhead. In recent years, the "one-shot" bender has gained much popularity because of the speed and ease it adds by making bends with just one "shot" or stroke of the bending ram against the conduit in the bending shoe. This type of bender provides great economy in bending, in many cases proving the difference between use of field bends or manufactured elbows.

Pipe Pushers are hydraulically powered ram assemblies used to push conduit through earth. They are used for installing underground conduit without digging a ditch the length of the conduit run. Only a working hole must be dug to provide positioning of the pusher. This technique of conduit installation is used where conduit is to be installed under existing concrete —such as a sidewalk, a driveway, etc. A typical pushing job used a gasoline-engine driven power pump for the hydraulic pressure to push 45 feet of 3½-in. conduit under a 26-ft. wide city street.

Cable Tools

Pulling Equipment used to pull wires and cables into conduit and other raceways varies widely in type and size to meet the requirements of various jobs. Such devices can range from the conventional fish-tape pullers for relatively small circuit work to large manual and power winches for pulling large cables or for long or difficult pulls on groups of small cables. One-man operation of small hand-held unit power pullers provides fast efficient wiring of conduit on much branch circuit and small feeder work. Other power pullers consist of truck or platform mounted units made up of large, high power winches with adjustable-speed drive, adjustable boom and pulley attachments or other features to provide a strong pull with adequate bracing of the pulling rig. Selection criteria for a power winch application are pulling force, portability and ease of setup. Power for the smaller, portable cable pullers is generally provided by electric motors with ready power cord connection to an outlet. The larger capacity pullers and power winches are commonly driven by gasoline engines. These larger units often have four-wheel bases for ready mobility and adjustable capstans for any direction or angle of pull.

Cable Reels of many types are used in modern electrical work to facilitate many steps in the payout of wire and cable into conduits. Many contractors provide special reeling of the proper number and lengths of conductors for

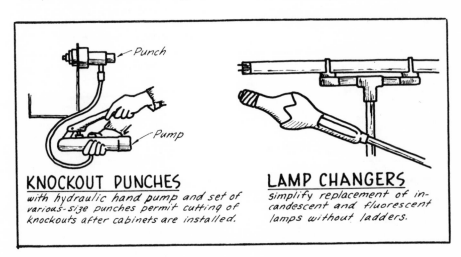

KNOCKOUT PUNCHES
with hydraulic hand pump and set of various-size punches permit cutting of knockouts after cabinets are installed.

LAMP CHANGERS
simplify replacement of incandescent and fluorescent lamps without ladders.

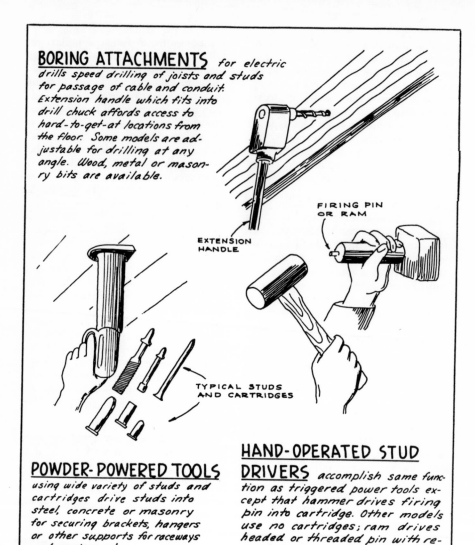

BORING ATTACHMENTS *for electric drills speed drilling of joists and studs for passage of cable and conduit. Extension handle which fits into drill chuck affords access to hard-to-get-at locations from the floor. Some models are adjustable for drilling at any angle. Wood, metal or masonry bits are available.*

EXTENSION HANDLE

FIRING PIN OR RAM

TYPICAL STUDS AND CARTRIDGES

POWDER-POWERED TOOLS *using wide variety of studs and cartridges drive studs into steel, concrete or masonry for securing brackets, hangers or other supports for raceways and equipment.*

HAND-OPERATED STUD DRIVERS *accomplish same function as triggered power tools except that hammer drives firing pin into cartridge. Other models use no cartridges; ram drives headed or threaded pin with repeated blows of hammer.*

specific jobs. Such pre-work preparation of wire and cable can add real speed and efficiency on the job. Other cable reel units combine a number of individual reels on a large rolling frame for selection of various sizes and types of wires to meet varying job requirements.

A NUMBER OF SPECIALTY devices are used in the installation of electric wires and cables, including the following:

Pulleys and rollers—A variety of pulleys and frame-mounted rollers are available for use in facilitating the pulling of cables into conduits in buildings or in manholes for underground work and for pulling interlocked armor cable into place on racks or trays where there are turns in the overall run. In a typical application, a piece of pipe is arranged to roll in a frame to permit cable to pass easily over it from the reel into the conduit which might be at a sharp angle with respect to the feed from the reel. Pulleys are commonly used by securing them to the rack for armor cable and pulling the cable through them, especially at turns in the run. They eliminate the friction which would be developed if the cable were pulled around non-rolling supports. Required pulling force is therefore greatly reduced by the use of such devices.

Connector crimpers—These tools are made in a variety of types and sizes to provide the force required to crimp connectors and splicing devices on cables. Some such tools are simple hand units of the pliers type. Others are hydraulically powered through hoses from hand- or foot-operated pumps. Similar crimp devices, commonly called indenters, are also used for indenter-type couplings and connectors for thinwall conduit.

Typical tools for fastening and supporting electrical equipment include the following:

Hand set with concrete nails is commonly used for fastening enclosures to concrete. A typical application is the securing of header duct to precast cellular concrete floor construction.

Powder-actuated stud driver is a very popular type of device for fastening equipment. It is like a pistol and uses a gunpowder cartridge. This fast, versatile and efficient tool can be used to drive mounting studs in steel and in a number of types of masonry construction. Light in weight and easy to handle in almost any position, this hand device takes selected cartridges to provide power loads for light, medium or heavy duty fastening jobs. Various types of studs can be driven, with threaded and non-threaded shafts for many types of fastenings.

Electric drills of many sizes, capacities and speeds are available to meet the range of job requirements for drilling wood, steel or other materials. Attachments extend the usefulness of regular electric drills. Right angle heads with speed changing features permit drilling in limited spaces. Metal-cutting hole saws, circular saws and reciprocating saw blade attachments provide for a very wide range of cutting operations. And bench stands can convert drills into handy drill presses for fabrication. Use of proper drill fittings makes possible cutting even masonry.

Other power tools in common use on modern electrical construction work include—electric hammers, electric impact tools, electric nutrunners for fastening nuts on bolts and electric screwdrivers. Electric hammers are made in two designs, one with reciprocating motion and the other with a mechanism to convert rotary motor

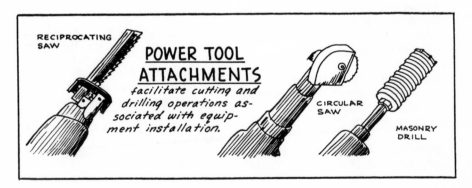

RECIPROCATING SAW

POWER TOOL ATTACHMENTS *facilitate cutting and drilling operations associated with equipment installation.*

CIRCULAR SAW

MASONRY DRILL

output into impact blows. Impact tools and nutrunners have high-torque rotary action for such tasks as tapping, nutrunning and screw driving.

Typical tools used for assembly phases of electrical work include:

Electric sanders are commonly used for cleaning bus bars to assure sound, low-resistance connections or to assure effective silver plating of the bars, when that is done prior to bolting connections.

Surveyor's transit and rod has become a standard contractor tool for leveling underfloor duct runs and header ducts.

Concrete drills are coming into common use by electrical contractors. On new construction, a typical use is for cutting necessary openings in cellular concrete floors for header duct feeds and for outlets. Many uses of such drills can be found on modernization projects to mount equipment in existing slabs or to carry circuits through slabs. Such units use diamond-tipped core drills.

Vacuum cleaners of the industrial type find application in cleaning out electrical enclosures and all types of drilled openings.

Scaffolding and platforms for overhead electrical work include the following types:

Telescoping booms on platform trucks for raising mechanics on work platform up as much as 40 feet to work on outdoor lighting fixtures.

Lift trucks with rising work platforms for overhead installations or maintenance. Typical use is in installation of lighting fixtures.

Pipe-frame scaffolds of many types and sizes for providing an elevated work platform. Assemblies are adjustable, with rollers for ready mobility or of the stationary type.

Aerial towers are hydraulically powered folding booms for elevating work platforms as high as 95 feet. Typical tower is mounted on back of a truck, with control of the elevation from the work platform.

Outside electrical work is greatly facilitated by tools and equipment used for both pole line and underground work.

Trenchers of all types are power driven machines which quickly and neatly dig trenches for ready installation of underground conduit or direct-burial cables. Typical units are self-propelled or hand-propelled and dig a trench from 3 to 8 inches in width (adjustable as required), up to 5 feet deep and at rates up to 17 feet per minute, depending upon the soil.

Bulldozers provide ready backfill of underground work with great speed and efficiency. Small, portable units are available.

Line trucks are made in a number of sizes and types for handling various phases of pole line work. Typical trucks may have booms or hydraulic cranes for setting poles or erecting light standards. Other trucks are equipped with earth digging augers for making pole holes. Pole trailers are still another type of rolling stock, used to transport poles to job sites.

A very wide variety of small hand tools are also used in pole line work. Special insulating gloves are one very important item. Others: protective shields for hot lines, pole-climbing spikes, safety belts.

Meters and Testers

A WIDE RANGE of electrical measuring and testing equipment is used in the construction, operation and maintenance of electrical systems and utilization devices. These include meters and testers for both ac and dc circuits, used either as hand-held instruments or panel-mounted instruments.

Although the terms "meter" and "tester" are frequently used interchangeably, precise use of each word can be based on the following definitions:

METER—An instrument which indicates a quantitative characteristic of an electrical condition, a measurement of actual unit values for a particular circuit—such as so many volts, or amperes, or watts, or the value of frequency or power factor.

TESTER—An instrument which indicates a qualitative characteristic of an electrical circuit. A continuity tester indicates that a path of current flow does or does not exist. A neon voltage tester indicates the presence or absence of a voltage across a set of terminals. Similarly, other testers provide a "yes" or "no" answer to investigations of electrical conditions—without actually measuring the value of the condition.

AC Meters

Instruments used for measuring the characteristics of alternating current circuits vary widely in complexity, depending upon the purpose for which they are used. Typical ac meters are as follows:

Voltmeters — These are electromagnetically operated units which are connected across circuits to measure the value of voltage. Internally, sets of coils (or a coil and iron vanes) are arranged to produce a form of motor action in response to current flow through the coils. And because the

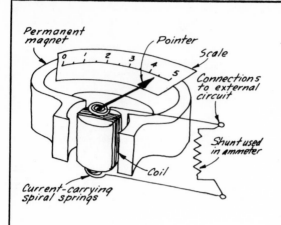

Permanent magnet — Pointer — Scale

0 1 2 3 4 5

Connections to external circuit

Shunt used in ammeter

Coil

Current-carrying spiral springs

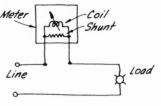

Meter — Coil — Shunt

Line — Load

MOVING COIL METER

Rectangular coil is suspended between poles of permanent magnet. When current from external circuit passes through coil, interaction of magnetic fields causes coil to turn on pivots. Pointer fixed to coil moves across calibrated scale. When current is removed, springs return pointer to rest at zero position.

AMMETER

The moving coil meter used to measure current is connected in series with the line. Since the amount of current which can be permitted to flow through the coil is limited, a shunt is used as shown to permit the measurement of higher currents. The current from the external circuit divides, part through the coil, part through the shunt. The portion through the coil is proportional to the total current.

current flow and its resultant torque are directly related to the voltage present at the meter terminals, the turning motion can be imparted to a spring-retained pointer to indicate voltage value on an appropriately calibrated scale of values. Such a scale is placed in the viewing window of all meters, with the pointer arranged to move across the scale to indicate values.

Calibration refers to the adjustment of a meter so that its reading is accurate. This is done in the factory initially against a known standard of voltage values. Some provision is usually made in the meter unit itself to permit setting of the pointer to the scale zero, to accommodate variation in spring tensions or other mechanical changes which vary the at-rest position of the pointer.

Voltmeters are contained in small assembly cases, with the viewing window on the front. In operating principle, there is no difference between hand-held meters, meters in combination volt-amp-ohm assemblies and meters mounted on control panels, on switchboards or on unit substation housings.

In use, voltmeters of the hand-held type are equipped with wire leads with clips on their ends. The clips are used to attach the leads to the terminals or conductors across which it is desired to measure the voltage. Switchboard meters are wired permanently into the bus conductors within the housing.

Ammeters—These measuring instruments operate on the same electromagnetic principles as the voltmeter.

But whereas there are two specific techniques suitable for voltmeter operation — electrodynamometer (fixed and moving coils) and iron-vane type (one coil with induced magnetic interaction between it and soft iron vanes) —only the latter type of meter movement, the iron-vane type is used for ammeters.

The basic ammeter of the type described above has a coil wound from a few turns of larger size wire rather than many turns of fine wire as used in the voltmeter of the same type. This ammeter is connected in series with the conductor and load to measure the load current in a circuit. For this type of hand ammeter, the circuit must be opened to permit insertion of the meter.

Switchboard or panel mounted meters may be directly connected into the circuit for which they will indicate current value or they may be coupled to the circuit through a current transformer when the current to be measured is too large for the meter.

This transformer uses the conductor being metered as its primary and produces a secondary current which bears a fixed ratio to the actual primary current. The ammeter is connected in the secondary circuit and is calibrated to provide scale indications of the actual current flowing through the circuit being metered.

A popular type of ammeter used for spot checking and testing of current flow through all types of electric conductors is the **clamp-on ammeter.** This is a hand-held, readily carried device which does not require opening of the circuit in order to measure current flow.

This type of ammeter actually consists of a current transformer and meter movement combined in a small tool-like device. At one end, curved clamp jaws are arranged to permit opening them to hook on to a conductor in which it is desired to measure the current flow. The clamp jaws are then closed so the conductor, in effect, passes through a closed loop at the end of the meter case.

In this position, the conductor under test acts as the primary winding of a transformer, with the jaws acting as a closed magnetic path to provide

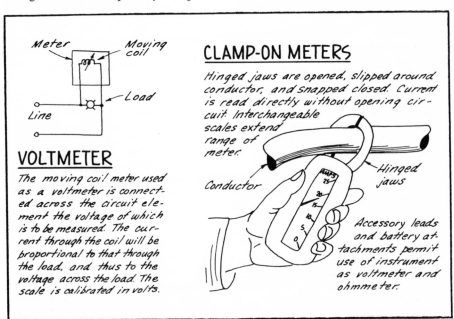

VOLTMETER

The moving coil meter used as a voltmeter is connected across the circuit element the voltage of which is to be measured. The current through the coil will be proportional to that through the load, and thus to the voltage across the load. The scale is calibrated in volts.

CLAMP-ON METERS

Hinged jaws are opened, slipped around conductor, and snapped closed. Current is read directly without opening circuit. Interchangeable scales extend range of meter.

Accessory leads and battery attachments permit use of instrument as voltmeter and ohmmeter.

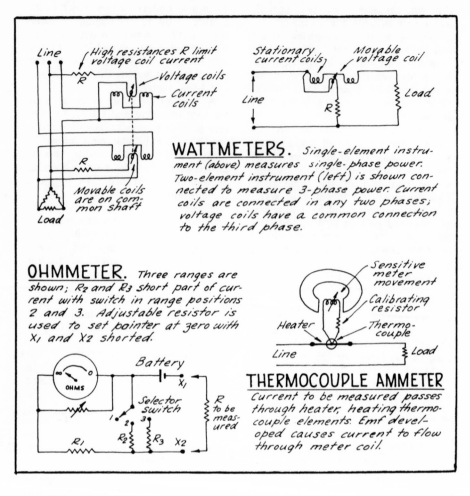

WATTMETERS. *Single-element instrument (above) measures single-phase power. Two-element instrument (left) is shown connected to measure 3-phase power. Current coils are connected in any two phases; voltage coils have a common connection to the third phase.*

OHMMETER. *Three ranges are shown; R_2 and R_3 short part of current with switch in range positions 2 and 3. Adjustable resistor is used to set pointer at zero with X_1 and X_2 shorted.*

THERMOCOUPLE AMMETER
Current to be measured passes through heater, heating thermocouple elements. Emf developed causes current to flow through meter coil.

an induced emf and current in the secondary coil within the meter case. This current operates the calibrated meter movement to indicate the value of current in the conductor being checked.

Clamp-on (or hook-on or snap-around) type meters are made in combination volt-ammeter units. Such units have one scale calibrated for reading amperes of current flow when used as described above.

They have another scale calibrated in volts for reading measurements made by means of wire leads which are connected across terminals to be checked. Some meters of this type are also equipped with a battery for check-

ing resistance. They have a scale calibrated in ohms, are used to test leads.

Clamp-on volt-ammeters are made in a range of sizes for circuits of low, medium and high current ratings, and for a wide range of voltage scales. A typical model might have amp ranges of 0-10/30/100/300/800 and volt ranges of 01150/300/750.

ALTERNATING CURRENTS and voltages may be measured by means of dc type permanent magnet, moving coil meters in combination with either a thermocouple or rectifier circuit.

Thermocouple Instruments—This type of metering device operates according to a phenomena known as the Seebeck effect: if the junction of two wires of dissimilar metals is heated, a small voltage is developed between the other two ends of the wires. In thermocouple instruments, a small resistance wire is provided to carry the current of the circuit being measured. This wire is in contact with the junction of the dissimilar thermocouple wires and produces heat at the junction in proportion to the current being measured. This heat then produces the thermocouple electromotive force, as previously described, which is connected to a very sensitive meter movement. The meter then registers the effect on a suitably calibrated scale, indicating either voltage or amperage. Instruments of this type are commonly used for measurements in radio-frequency and other high-frequency circuits.

Rectifier Instruments—The basic dc meter movement consists of a horseshoe (or "C" shaped) magnet with a pivoted movable coil mounted in the gap between the poles. A pointer is attached to the rotating coil assembly, which is spring-held at the zero position on the meter scale. In the presence of the fixed magnetic field, current through the coil will cause the coil assembly to rotate, like a dc motor, against the tension of the springs. The pointer registers the value of current (or voltage) on the meter scale. To use such a meter for measuring ac conditions, small dry type rectifiers are arranged in a bridge connection to provide full-wave rectification of the ac input. The dc output of the rectifier is then supplied to the dc meter. By means of the fixed relation between average and rms values of sine waves, such instruments can be calibrated for directly indicating the ac current or voltage. Rectifier instruments are commonly used for testing and servicing of equipment and appliances.

Wattmeters — Measurement of electric power (which is the time rate of electric energy consumption) in ac circuits is generally made by means of a wattmeter. In dc circuits, the power in a circuit is equal to the product of the current and the voltage of the circuit at any time. An ammeter and voltmeter can therefore be used to determine power in dc circuits. And these two meters can also be used to determine the power in ac circuits—but only when the power

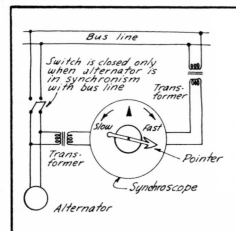

SYNCHROSCOPE, shown in simplified single-phase circuit, indicates relative difference in frequency between bus line and alternator, which is to be thrown on the bus.

Pointer rotates in one direction or the other as long as alternator is out of synchronism with bus line. Alternator is in synchronism when pointer remains motionless vertically at mark on dial.

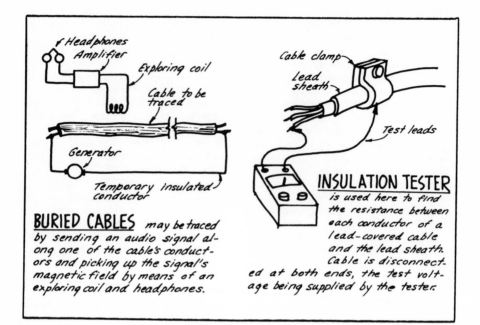

Headphones
Amplifier
Exploring coil
Cable to be traced
Generator
Temporary Insulated conductor

BURIED CABLES *may be traced by sending an audio signal along one of the cable's conductors and picking up the signal's magnetic field by means of an exploring coil and headphones.*

Cable clamp
Lead sheath
Test leads

INSULATION TESTER *is used here to find the resistance between each conductor of a lead-covered cable and the lead sheath. Cable is disconnected at both ends, the test voltage being supplied by the tester.*

factor is unity. When the power factor of an ac circuit is other than unity, the product of the circuit voltage and current does not represent the power of the circuit. The wattmeter measures power in ac circuits for any and all conditions of power factor.

Typical wattmeters are made in portable cases and for switchboard or panel mounting. Internally, they contain two sets of interacting coils, one fixed and one movable. One coil, called the current coil, carries the load current of the circuit being measured. The other coil is connected in a series with a resistance across the circuit load. With currents flowing through these coils, they act to produce a turning force on the movable coil assembly. This torque is proportional to the average power of the circuit for any power factor. It is the effect of the instantaneous product of current and voltage, varying as the power factor varies.

A wattmeter is connected into a circuit by connecting its current terminals in series with the load and its voltage terminals across the circuit. Wattmeters are rated in terms of the

current their series coils can carry and the voltage their potential circuits can withstand. In the application of wattmeters, special care must be taken to assure that the ratings of the internal coil elements are never exceeded.

Wattmeters are made in both single-phase and polyphase types. Power in three-phase circuits may be measured by two or three single-phase wattmeters or a polyphase wattmeter.

Watthour Meters — These meters measure the amount of electric energy consumed. There are single-phase types and three-phase types. Their internal construction contains a motor-like movement which drives clock-dial or cyclometer registers to indicate the sum of kilowatthours of energy consumed.

The most common and most familiar watthour meter is the residential meter at the service entrance location in every home. Such meters may be of the self-contained type with provisions for line and load connections or of the socket type for plugging into prewired meter sockets, either individual socket housings or through housings for multiple meters.

Power Factor Meters—These are instruments commonly used in industrial applications to measure the power factor of feeders, load areas or individual load devices. These meters operate on the interaction between fixed and movable coils, in the same general manner as other electrodynamometer instruments. Both single-phase and three-phase meters are made. For use on three-phase circuits, the meter indicates power factor (or it can be calibrated in degrees of pf angle) for essentially balanced conditions only. The most common type of pf meter is the switchboard mounting type.

Var Meter—As its name suggests, this meter is used to measure and indicate the amount of *volt-amperes reactive* (var) in a circuit. The three capacity criteria of an ac circuit are: watts, or effective power; volt-amperes, the product of voltage and current, which is greater than watts when the power factor is not unity; and reactive volt-amperes, which is a measure of the corrective volt-amperes, either leading or lagging, needed to produce unity power factor in the circuit.

Synchroscope — This instrument is similar in operation to the power factor meter. It is used to indicate when generators are in proper phase relation for connecting in parallel.

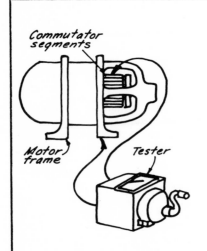

MOTOR INSULATION *may*
be tested with a "Megger" or similar insulation resistance tester. Resistance is measured between motor frame and each commutator segment, representing the ends of each commutator coil. Tester gives readings in megohms.

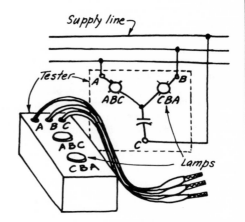

PHASE SEQUENCE INDICATORS
establish order in which each phase of a 3-phase line passes through its peak value. With test leads connected to supply line as shown, one lamp will glow more brightly than the other, indicating either an A-B-C or C-B-A sequence. The capacitor is included to equalize the impedance between phases.

ELECTRICAL TEST instruments range from simple devices which determine the presence or absence of an electrical potential to complicated test sets which measure conditions of operation or detect, locate and indicate the nature of all types of faults in electrical circuits and machines.

Standby of the old-time electrician for use on 115/230-volt 3-wire circuits was the single 230-volt incandescent lamp in a pigtail socket with two leads attached. Across a 230-volt line the lamp burned brightly; a dim light indicated 115 volts. No response meant either a dead circuit or a burned-out bulb.

Today's more-reliable pocket voltage tester is calibrated to indicate a wide range of voltages and is as much a part of the maintenance man as his screwdriver.

Insulation Testers. Among other things, heat, dirt and moisture cause a gradual deterioration of the materials used for insulating electrical conductors and equipment. A wide variety of test instruments is available to the electrician for his troubleshooting and routine preventive maintenance as well as to the equipment repair shop and manufacturer to check the effectiveness of insulation. Most of these instruments operate on the principle that even insulation conducts minute currents through it when a voltage is impressed across it, the magnitude of the current depending upon the magnitude of the voltage and the condition of the insulation. The instrument converts the effect of the voltage and current into megohm readings on its scales. All equipment and conductors are de-energized prior to being tested, the test voltage being supplied by the tester itself.

The familiar Megger incorporates a dc generator, either hand-operated by means of a built-in crank or motor-driven. Such instruments have ranges up to 50,000 megohms and 2500 volts. Models are also available with a built-in rectifier, being plugged into a 115-volt general-purpose circuit during use.

Another similar insulation tester uses a built-in battery to furnish a small dc voltage which is changed to ac by a vibrator, stepped up to high voltage by a transformer, and then rectified to provide the desired dc test voltage at the terminals of the set.

For purposes of maintenance, records of insulation tests are kept for each individual piece of equipment or cable. If successive tests are made under similar conditions of moisture and cleanliness, the test readings will give a fairly accurate indication of the rate of deterioration of the insulation or may point out a source of trouble.

Ground Testers are similar in construction and operation to insulation testers, being used to measure resistance to earth of equipment ground connections, except that lower values of resistance are involved. Typical scales range from 0 to 3 to as high as 0 to 30,000 ohms.

Growlers. The armatures and stators of rewound motors and generators are tested for faults using a testing magnet called a growler. This is essentially a U-shaped core made up of thin iron sheets wound with a coil of wire. When placed against the armature over the armature coils, the growler acts like a transformer, with the armature coils as the secondary. The action of a thin iron strip, or feeler, which is either built into the growler or held by the operator against adjacent armature slots, indicates the existence of short circuits in the windings. The poles of the growler core are adjustable to suit the size and coil spread of the motor or generator being tested. By means of selector switches, the growler is also used to test for open circuits and grounds.

Rotation Testers. This instrument provides a means of assuring correct rotation of a polyphase motor when it is energized to prevent damage to driven equipment and to eliminate temporary connections or reconnection of leads. The tester leads are connected to the motor leads and the motor is rotated by hand in the correct direction. By observing the deflection of the meter needle, the operator can tell whether the phases are connected in A-B-C sequence. The leads can then be tagged and connected to the supply line in the same order.

Phase Sequence Indicators. All feeders and branch circuits in a distribution system must be properly interconnected (phase A to phase A, etc) throughout to prevent damage to or improper operation of connected equipment. Indicating sets are used to determine the phase relationship of any energized 3-phase line. They usually consist of two light bulbs and a condenser connected in a "Y" configuration, with a test lead connected to each leg of the Y. The two lamps are labeled A-B-C and C-B-A, the two possible phase sequences. When the three leads are connected to a three-phase line, one lamp will burn more brightly than the other. The bright light indicates the phase sequence.

Cable Fault Locators. Underground cables which develop grounds, shorts or opens may be repaired without complete unearthing by first locating the exact spot of the fault by means of suitable test equipment and then digging at that spot in the case of buried cable or, where the cable is run in ducts, by pulling the cable into the manhole or vault nearest to the fault.

Bridge circuits such as the Murray loop, which is a modification of the well known Wheatstone bridge, are incorporated into testers which provide a means of calculating the distance to a fault along the length of the cable. Where cable routes and lengths are not accurately recorded, fault locators using a signal generator and explorer coil may be used. Exact procedures differ depending upon the type of fault; however, the principle in each involves sending a signal along a conductor in the faulted cable and then following the cable with an "exploring" or "pick-up" coil connected to headphones through a suitable audio amplifier. A change in the nature of the tone received or its sudden absence indicates the location of the fault. This equipment may be used on cables buried or exposed and is frequently used to trace the route of a buried cable.

Pole Line

Equipment

WITH CONTINUING expansion of power generating, transmission and distribution systems throughout the country, the volume of outside electrical construction work continues to grow.

In particular, pole line construction is increasing to provide distribution facilities necessary to transport bulk electrical energy to old and new centers of electrical utilization.

Unprecedented growth in the consumption of electrical energy is demanding more and longer lines of distribution. All of this adds up to more public pole lines. And the general growth in size of modern industrial plants—with multiple-building layouts of large single-story buildings served by many load-center substations—has stimulated construction of private pole lines.

From the standpoint of equipment, all pole line work involves a very wide range of types and sizes of materials, apparatus and tools.

Wood Poles

A number of different types of wood are used for wood poles. Typical species are: northern white cedar, western red cedar, western firs, douglas fir, western hemlock, northern pines, southern pines and western pines.

Specifications and dimensions for wood poles have been standardized by the American Standards Association. Typical spec requirements are as follows:

1. Certain defects in the poles are generally prohibited: cross breaks (cracks), bird holes, plugged holes, hollow butts or tops, damage by marine borers, splits or through-checks in the top, decay and the presence of nails, spikes or other metal not specifically authorized by the purchaser.

2. Certain defects are generally permitted: sap stain that is not accompanied by softening or decay of the

CLASSIFICATION OF WOOD POLES

Poles are divided into classes 1 through 10.

In general, the lower the class number, the larger the pole circumference.

Each class has a minimum circumference, measured at top of pole:

CLASS	1	2	3	4	5	6	7	8	9	10
MINIMUM TOP CIRCUMFERENCE (INCHES)	27	25	23	21	19	17	15	18	15	12

Different types of poles must meet minimum circumference requirements 6 feet from the butt end to qualify for a specific class designation.

Typical requirements for 30-ft poles:

TYPE OF WOOD	CLASS									
	1	2	3	4	5	6	7	8	9	10
	MIN. CIRCUMFERENCE 6 FT. FROM BUTT (IN.)									
NORTHERN WHITE CEDAR	47½	44½	41½	38½	35½	33	30½	NO		
WESTERN RED CEDAR	41	38½	35½	33	30½	28½	26½	B U T T		
CREOSOTED SOUTHERN PINE	37½	35	32½	30	28	26	24	RE QUI RE-		
CHESTNUT	40	37½	35	32½	30	28	26	M EN TS		

COMMON POLE-SETTING DEPTHS (FEET)*

Length of pole	Depth in soil	Depth in rock
20	5	3
30	5.5	3.5
40	6	4
50	7	4.5
60	7.5	5
70	8	6
80	9	6.5

*Add 6 inches for poles on curves or corners.

EXAMPLE:
A 30-ft Northern White Cedar, 27 in. top circumference, would be Class 1 only if circumference 6 ft from butt end were at least 47½ in. But a Creosoted Southern Pine would be Class 1 with only a 37½-in. circumference at that point.

Similar requirements exist for poles of other lengths.

POLE SETTING by hand

requires cooperation and teamwork of entire crew plus special tools.

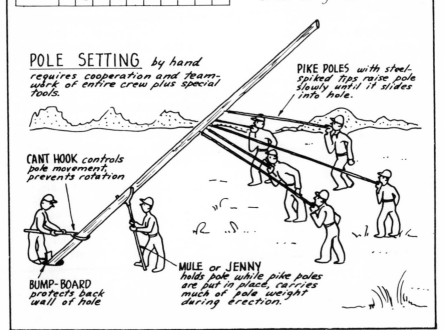

PIKE POLES with steel-spiked tips raise pole slowly until it slides into hole.

CANT HOOK controls pole movement, prevents rotation

BUMP-BOARD protects back wall of hole

MULE or JENNY holds pole while pike poles are put in place, carries much of pole weight during erection.

wood, a small amount of twist of grain (spiral grain) along the length of the pole, hollow pith centers in tops, butts and knots of poles to be treated for their full length.

3. Certain defects are permitted to a limited extent: checks and splits in butts, insect damage, knots, defective butts due to splinter-pulling in felling the tree, slight curvature of the pole along its length.

4. Poles less than 50-ft in length shall be not more than 3 inches shorter or 6 inches longer than nominal length. Poles over 50-ft long shall be not more than 6 inches shorter or 12 inches longer than nominal length.

● **Pole Classifications**—Wood poles are classified according to the circumference at the top of the pole and the circumference at a point 6 feet from the butt.

There are ten classes, designated Class 1, Class 2, etc. to Class 10. For each class, there is a minimum top circumference and a minimum bottom circumference which depends upon the length of the pole. Typical poles vary in length from 16 to 100 feet. Pole dimensions for each class also vary for different types of wood used for the poles.

● **Decay Prevention** — To assure full strength and long life of poles, they are generally impregnated with a substance which will resist and retard decay of the wood.

Southern pine poles and Douglas fir poles are commonly treated for their full length to afford maximum protection. Other woods require only treatment of the butts of poles to a point about one or two feet above the ground line when the pole is set.

A number of different preservatives and treating methods are used on wood poles. On all poles, all outer bark is removed after the poles have been seasoned for a number of months.

Inner bark must also be removed in accordance with various spec requirements, depending upon the type of treatment to be given to the wood. All poles are neatly sawed at the top and at the butt, and overgrown knots and branch stubs are trimmed close.

Poles to be full treated are then roofed, gained (cut for mounting of crossarms) and bored (for bolt holes). One type of treatment consists of submerging the poles in a hot coal-tar derivative for at least 15 minutes.

Another treatment involves a hot creosote bath for 15 minutes; or, in a more extensive treatment, a long hot creosote bath is followed by a long cold creosote bath.

For butt treatment, the poles are suspended with their butts submerged in the treating substance up to a point 1½ feet above the ground line.

For full length treatment to assure maximum preservation of any species of wood, pressure processes are used to provide overall positive impregnation, minimizing decay and adding life to poles in service.

● **Coded Brand** — Standard specifications for the manufacture of wood poles require that the poles be stamped or branded legibly with coded nota·tions to provide certain basic data on the pole: the code or trademark of the supplier, the plant location and the year of treatment, the species of wood used for the poles and the preservative used, a numeral indicating the class of the pole and numerals giving the length of the pole.

● **Pole Storage**—Whenever poles are stored, they must be stacked on treated skids in such a way as to support the poles without producing noticeable bending or distortion in any of them. Usually creosoted or other decay resisting skids are used for storing poles.

The poles are commonly piled to permit free circulation of air through them, and the bottom poles in the pile are at least one foot above the general ground level or any vegetation growing there. And poles should not be piled too high to avoid damage to the bottom poles.

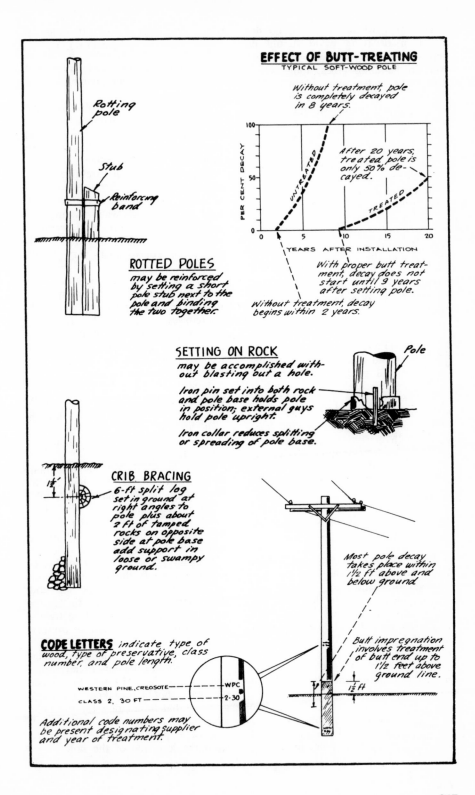

ROTTED POLES

may be reinforced by setting a short pole stub next to the pole and binding the two together.

EFFECT OF BUTT-TREATING
TYPICAL SOFT-WOOD POLE

Without treatment, pole is completely decayed in 8 years.

After 20 years, treated pole is only 50% decayed.

With proper butt treatment, decay does not start until 9 years after setting pole.

Without treatment, decay begins within 2 years.

SETTING ON ROCK

may be accomplished without blasting out a hole.

Iron pin set into both rock and pole base holds pole in position; external guys hold pole upright.

Iron collar reduces splitting or spreading of pole base.

CRIB BRACING

6-ft split log set in ground at right angles to pole plus about 2 ft of tamped rocks on opposite side at pole base add support in loose or swampy ground.

Most pole decay takes place within 1½ ft above and below ground

Butt impregnation involves treatment of butt end up to 1½ feet above ground line.

CODE LETTERS indicate type of wood, type of preservative, class number, and pole length.

WESTERN PINE, CREOSOTE — WPC
CLASS 2, 30 FT — 2·30

Additional code numbers may be present designating supplier and year of treatment.

To safeguard stored poles, no decayed or decaying wood should be permitted to remain underneath them. And in handling poles, tools which will make indentations deeper than one inch—pole tongs, cant hooks and other pointed tools—should not be used, particularly on the ground line section of poles. Treated poles should not be dragged along the ground.

For a particular job, the height of poles to be used depends on many conditions. For one thing, local municipal ordinances often regulate height of poles and methods of stringing wires. Common practice dictates certain minimum spacings between ground and the lowest line wire for various conditions: 18- to 21-ft. for highways, 15-ft. above sidewalks and either above trees or through lower branches.

IN ADDITION to clearance over buildings and other obstructions, the number and types of crossarms to be mounted on each pole determines the height of poles to be used—the more crossarms, the higher the poles.

And spacing of poles in pole lines varies with particular conditions of an installation. Common pole spacings for typical industrial and utility lines are between 100 feet and 150 feet, although much greater spacings —up to 500 feet—have been used on lines using ACSR (aluminum cable, steel reinforced).

• **Pole Setting**—Setting of poles varies according to the type and size of the poles and the conditions of the ground in which they are set. One rule of thumb has it that poles in straight runs should be set in normally solid ground to at least one-sixth of their length.

Typical depths for pole settings in normally firm ground are: 5½ feet for 30-ft poles in straight-line pole runs; 6 feet for 30-ft poles in line curves, corners or points of extra strain; 7 feet for 50-ft poles in straight runs; 8 feet for 60-ft poles in straight runs; 9½ feet for 70-ft poles at curves, corners and strain points.

Typical pole line installation procedures and requirements are as follows:

In general, all the poles in a run should be set to maintain an even grade. When the average run is level, consecutive poles should not vary more than 5 feet in height.

In a pole line run careful selection of the locations for setting the poles— avoiding setting poles in extremely high or low points when passing over uneven but generally level terrain— can assure relatively level lines without using poles which vary greatly in length.

In straight pole line runs, all poles should generally be set perpendicular to the ground. However, where the line curves or angles, poles are usually set with a slight slant away from the direction of pull which will be exerted by tension of the installed conductors.

Pole holes must be large enough to accommodate the pole butts without need for slicing or shaving the pole and to permit tamping of the dirt and fill in the hole around the pole after it is set.

Such holes should be essentially constant in diameter for their depths. And after the hole has been refilled and tamped solidly, dirt should be piled in a cone around the butt of the pole to provide drainage of the water away from the pole.

Whenever it is necessary to shorten a pole on a job, this should be done by sawing a piece off the top end. And when this is done to treated poles, the bare end should be treated with a further application of preservative of the type used.

When poles are set along the edge of cuts or embankments or any other place where the soil might be washed out, care should be taken to provide a substantial foundation for the pole. On embankments, the depth of setting of a pole should be measured from the low side of the pole.

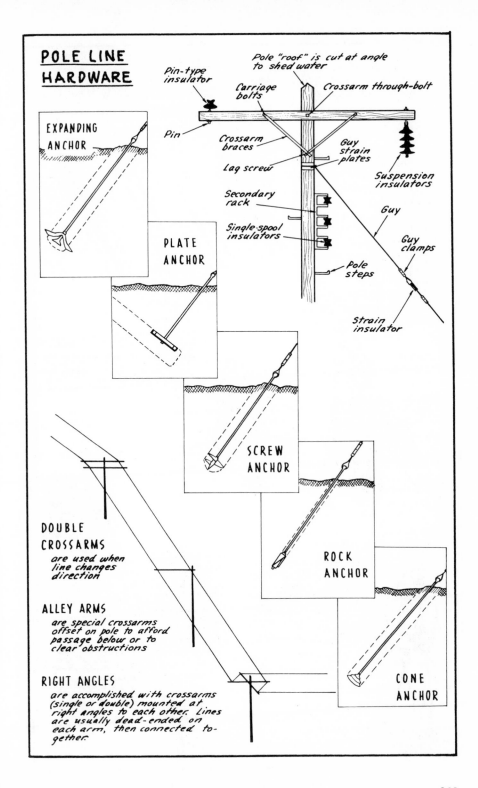

POLE LINE HARDWARE

EXPANDING ANCHOR

PLATE ANCHOR

SCREW ANCHOR

ROCK ANCHOR

CONE ANCHOR

Pin-type insulator

Pin

Pole "roof" is cut at angle to shed water

Carriage bolts

Crossarm through-bolt

Crossarm braces

Lag screw

Guy strain plates

Suspension insulators

Guy

Guy clamps

Secondary rack

Single-spool insulators

Pole steps

Strain insulator

DOUBLE CROSSARMS
are used when line changes direction

ALLEY ARMS
are special crossarms offset on pole to afford passage below or to clear obstructions

RIGHT ANGLES
are accomplished with crossarms (single or double) mounted at right angles to each other. Lines are usually dead-ended on each arm, then connected together.

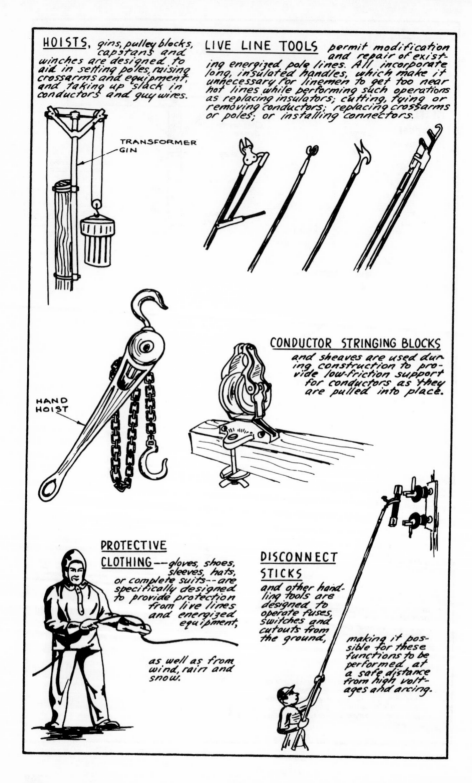

HOISTS, gins, pulley blocks, capstans and winches are designed to aid in setting poles, raising crossarms and equipment, and taking up slack in conductors and guy wires.

TRANSFORMER GIN

LIVE LINE TOOLS permit modification and repair of existing energized pole lines. All incorporate long, insulated handles, which make it unnecessary for linemen to get too near hot lines while performing such operations as replacing insulators; cutting, tying or removing conductors; replacing crossarms or poles; or installing connectors.

HAND HOIST

CONDUCTOR STRINGING BLOCKS and sheaves are used during construction to provide low-friction support for conductors as they are pulled into place.

PROTECTIVE CLOTHING—gloves, shoes, sleeves, hats, or complete suits—are specifically designed to provide protection from live lines and energized equipment,

as well as from wind, rain and snow.

DISCONNECT STICKS and other handling tools are designed to operate fuses, switches and cutouts from the ground,

making it possible for these functions to be performed at a safe distance from high voltages and arcing.

The methods and procedures for setting poles in other than normally firm ground vary widely with the type of ground and with general construction characteristics of the pole line itself.

Various special setting methods are used for poles in weak ground, sandy soil, swampy areas and in rocky ground, as follows:

Concrete is commonly used for setting important poles in loose ground.

In swampy grounds "sand barrel" setting is widely used. This consists of using strong steel cylinders or strong barrels (in either case about 3 feet long) to facilitate digging of the pole hole, pushing the barrels down to form the walls of the hole as it is dug. The barrels keep the ground forming the sides of the holes from collapsing during the digging.

In such cases, the pole is set in the barreled hold and the inside of the barrels are then filled with clay, sand or other firm soil.

If wooden barrels are used for this method, they may be left in place in the ground; if steel-cylinder type "sand barrels" are used, they are removed by hoisting them out of the ground after the pole is set and the hole filled.

In either case, the pole base is given a much larger bearing surface to provide greater support for it in the loose ground.

Another technique used to strengthen the setting of poles set in swampy grounds uses braces attached to the pole butt just above or beneath the ground level.

Such a brace is commonly triangular, with a cross piece attached to the pole butt about a foot underground and a supporting piece attached between each end of the cross piece and the pole at a point half way up the pole to the first cable on the line.

The pieces used for making such braces are usually lengths of old poles. Crossarm bolts are used to attach the pieces. Bracing methods are recommended wherever concrete foun-

dations would be used if the line were important.

Another pole setting technique used in marshy areas is the "A" structure. This consists of two poles set about 6 feet apart at the base but with their tops brought together and clamped, and with a common cross piece of an old pole attached between the butts of the two poles beneath the ground.

Setting poles in rock requires blasting of the pole holes. A typical procedure for such setting consists of first drilling holes about 18 inches deep at the point of the pole hole and the guy anchor positions. Then a stick of dynamite is set off in each hole.

Each hole is cleaned out after the blasts, and the procedure is repeated until the required pole and anchor hole depth is obtained. Holes in rock provide strong support for poles and do not have to be as deep as holes in normal ground.

● **Erecting Poles**—Modern pole setting is usually done with the aid of pole-setting trucks—line trucks fitted with booms and winches to lift and move the poles. In areas where there is no access or room for maneuvering vehicles, the poles are set by hand. Experienced line crews can upend the pole and slide it into the hole in a matter of minutes using sharp-pointed pike poles, "mule" or "jenny" pole rests, and cant hooks, used at the base to prevent rotation. Dirt and stone is then packed around the pole with tamping bars.

TWO IMPORTANT phases of pole line construction work are: (1) the assembly of the pole-top equipment for supporting the conductors, and (2) the guying of the overall structure. Details on this work cover the following equipment:

Crossarms

Wood crossarms are generally made of Douglas fir or treated southern yellow pine. Although there are a number of specifications for wood crossarms and although standard characteristics have not been established, crossarms in general use do not vary widely in dimensions for particular applications.

Typical crossarms will vary between 5 feet 7 inches and 10 feet in length, with cross sections between 2 by 3 inches and 4 by 5 inches, depending upon required conductor spacing and the number of insulator pins to be mounted on the crossarm.

Single crossarms are generally used on poles in straight line runs, with normal loads and normal spans. Double crossarms, one on each side of the pole, are used at angles in pole lines, where direction of lines change, to provide additional strength due to unidirectional forces exerted by tension of the conductors.

When poles are lined up for straight line runs, the crossarms are mounted at right angles to the direction of run. At points in a pole line where the direction of the line changes, crossarms on poles at such points are generally mounted to bisect the angle formed in the line.

When the angle in a pole line is greater than 60 degrees, the two sections of line coming into the angle are commonly dead ended—each section brought to an insulated support assembly.

In new construction, crossarms, braces and other fittings are commonly mounted on the poles before they are set in the ground. In changeover of lines on poles and in other alterations to pole lines, crossarms are mounted on poles which are already set.

In addition to wood crossarms, galvanized steel angle crossarms are also used on wood poles to obtain greater strength for supporting conductors— as on poles at which the conductors are dead ended. Steel crossarms are also more durable than wood crossarms.

Construction of pole lines requires the use of a wide range of sizes and types of accessory equipment and devices, as follows:

Insulators

A variety of types and sizes of insulators are used on pole assemblies to support the conductors. The most common are **pin type** and **suspension type insulators.** Size and shape of insulators depends upon circuit voltages.

Pin type insulators are made of either glass or porcelain, with a threaded hole through their centers. In use, they are screwed onto threaded insulator pins mounted on crossarms. Such insulators are made with grooves on the sides for low voltage applications, and with grooves on top for high voltage applications.

Insulator pins used to mount pin type insulators on crossarms may be made of wood or metal or combinations of wood and metal. In all cases, they are constructed to provide mounting on a crossarm and have a threaded end to accept the threaded pin insulators.

Suspension type insulators are used to support aerial conductors under crossarms instead of on top of crossarms. They are commonly used where voltage of a line is so high that pin type insulators would have to be too large to provide the required amount of insulation.

The top part of a suspension type insulator is commonly known as the bell and the bottom part—the large diameter dish part—is known as the petticoat. These insulators may be supported from the crossarm by U-shaped clamps known as "clevises" in the case of steel crossarms or by hooks in the case of wood crossarms.

Circuit conductors are supported from the bottom of suspension insulators by means of a clamp. In application, a number of these insulators may be arranged in series in a string. The greater the number of insulators in a string, the higher is the voltage rating of the assembly.

In addition to the above described insulators, there are a number of other types of insulators used in pole line work. One such insulator is the **spool insulator** which is widely used on insulator rack assemblies (also called secondary racks) mounted on sides of poles to support low voltage distribution circuits or to provide dead ending.

Another regularly used type of insulator is the **strain insulator** used in guy wire assemblies to insulate the lower section of the guy wire from the dangerous voltages on the pole top. Strain insulators are also used for dead ending.

Anchors

To provide strength to the support which guy lines give to poles, a firm purchase must be obtained in the ground. Anchors are devices used to provide solid attachment of guy wires to the ground. They are made in a number of types to provide different amounts of support and for use in various ground conditions.

Commonly used types are the (1) expanding, (2) plate, (3) screw, (4) rock, and (5) cone anchors.

• **Expanding anchors**—which are used for general anchoring applications. This type of anchor is a collapsible assembly of several sections of steel about 10-in long. It is inserted in the anchor hole, lowered to the bottom of the hole, then expanded by taps on a tamping bar to force the assembly to spread.

In its expanded position, the anchor spreads out to wedge itself into the ground on the sides of the hole. The anchor rod which connects to the guy wire above the ground is attached to the anchor before it is lowered to the bottom of the hole. The hole is backfilled with small stones and dirt, tightly tamped to pack the dirt.

• **Plate anchors**—which are also used for general anchoring applications. With this anchor, the anchor hole is dug at an angle perpendicular to the angle at which the anchor rod enters the ground. The plate anchor is, just as its name implies, a curved, rectangular plate of steel about 8 by 25-in in a typical case. This anchor is placed in the anchor hole and attached to the anchor rod which has been driven into the ground until it enters the anchor hole. The hole is then backfilled.

• **Screw type anchors**—which are quickly installed anchors used for general anchoring and for anchoring in sandy or damp soil where it would be difficult to dig an anchor hole. This anchor consists of a large diameter, short section of screw, which is screwed into the ground like a corkscrew, by means of a bar inserted through the eye of the anchor rod.

• **Rock anchors**—which are used to provide anchoring in rock or extremely rocky soil. In such cases, standard types of anchors could not be used. The rock anchor is installed in a hole drilled in the rock and develops holding power by self-wedging action.

• **Cone anchors**—which are used for anchoring in rock when the anchor hole is blown with dynamite or for anchoring in a hole dug in shale. This anchor is simply a steel cone to which the anchor rod attaches. It is placed in the hole and backfilled.

Tools and Equipment

Because pole line work is heavy construction, a wide range of tools and equipment is commonly used for various phases of the work. This includes transportation vehicles, pole handling and erecting equipment, hole digging equipment, many small tools for handling poles and tools and equipment used for wiring poles.

• **Line Trucks**—Many different types of trucks are used for line construction work. The largest types of trucks are called full-body line trucks. On these trucks the large body behind the cab consists of side walls which are divided on the outside into various sizes of compartments to hold materials, tools and equipment. These compartmented walls semi-enclose the body of the truck where the line crew sits and where large tools are stored. The top of the body can be enclosed by sliding covers which engage the sides. Full-body line trucks are commonly equipped with winch and derrick arrangements to handle setting of poles up to 75-ft long.

Smaller line trucks have compartmented bodies with shelves and bins to hold supplies and are equipped with racks to hold various pole sticks. These are similiar to standard pick-up trucks. They may also be equipped with extension ladders and /or with small capacity winch and derrick setups for poles up to 35-ft.

Many line trucks, both large and small, may be equipped with hydraulic-lift platforms. Such equipment can be mounted on the truck to provide a rising platform from which mechanics can work on pole mounted luminaires or other equipment.

● **Pole Trailers**—A number of different types and sizes of pole-hauling trailers and donkeys are made for attaching to the rear of pole line trucks to carry poles to the job. These are heavy duty, steel frame assemblies with two or four wheel carriages. The heavier units are called trailers; the lighter units—dinkeys.

On either type, provisions are made for supporting the poles and tying them down during transit. In the larger sizes of pole trailers, a variety of constructions are available with such features as: removable fenders; storage space for other equipment, such as pole transformers; power brakes on the wheels; telescoping connection between carriages of four-wheel trailers to adjust trailer length for different lengths of poles.

● **Pole Hole Diggers**—Earth-boring machines are used for digging pole holes and anchor holes. They are

available in several types for use on the rear of different types of pole line trucks. Operating just like big drills, hole diggers are usually arranged to swivel into any desired position and are under the control of the operator who can adjust speed of digging, direction of hole and depth of digging.

Common augers used can make holes from 9 inches in diameter up to 60 inches in diameter, up to 11 feet deep for poles up to 110 feet long. Pole diggers may be powered from the truck engine or from a separate engine—such as a diesel engine mounted on the back of a platform-body truck. The latter case is typical of combination hydraulic earth-boring machines and pole setters, widely used today.

● **Pole Setters**—Derricks commonly used for setting poles in their holes are tripod assemblies of tubular steel legs mounted on the rear of line trucks. Two of the tripod legs are stiff and the third, or middle leg, is of telescoping construction to permit adjusting the height of the top point of the rig where the three legs come together to hold the sheave over which the pole-lifting cable is passed from the winch on the floor of the truck to the lifting hook. With common tripod derricks, the position of the tripod is adjusted manually by means of the telescoping leg.

Such derricks are demountable to store in the rear of the truck. In the case of other types of derricks, one winch drum is used to raise and lower the derrick to desired position; a second winch drum is used to operate the lifting line.

● **Line Tools**—Many types of tools and equipment are used for work on hot line conductors.

Typical tools include: wire tongs and fittings, which consists of a length of insulating wood stick (8 to 14 feet) with fittings on the ends to permit handling and holding live conductors clear of wiring areas and to permit moving conductors to auxiliary arms; link sticks used to spread and hold conductors aside at mid span when moving a pole; auxiliary crossarms to provide change of insulators or crossarms; lever type wire cutters, which consist of a pair of wire cutters attached to the end of a 4-6-ft stick and operated by an insulated handle at the opposite end of the stick; sticks for handling insulators on hot lines; sticks for installation and removal of all types of hot line ties.

In addition to the wire assortment of hand tools for installing and maintaining pole top assemblies of hot conductors, various types of protective equipment are used.

Special static resisting coats are made to protect linemen from shock hazards in rainy or wet weather. Lineman's rubber gloves are available insulated for various high voltages. Insulating blankets and jackets of rubber are made for draping over hot conductors and equipment to prevent accidental contact by linemen during hot line work.

INDEX